MARK CATESBY'S LEGACY:
Natural History Then and Now

"As FIGURES convey the strongest Ideas, and determine the Subjects treated of in Natural History, the Want of which hath caused so great Uncertainty in the Knowledge of what the Antients have described barely by words; in order to avoid such Confusions, we shall take care to exhibit every thing drawn by the Life..."

MARK CATESBY
From his proposals for printing an
essay toward a natural history of
Florida, Carolina and the Bahama Islands

1729

M.J. Brush
Alan H. Brush
2018

The Catesby Commemorative Trust

© Published 2018 by The Catesby Commemorative Trust
www.catesbytrust.org
Written and Illustrated by Alan H. Brush and M.J. Brush
© Copyright Text by Alan H. Brush 2018
© Copyright Illustrations by M.J. Brush 2018
www.mjbrush.com
All rights reserved.

This book may not be produced, in whole or in part, including illustrations, in any form (beyond that copying permitted by Sections 107 and 108 of the U.S. Copyright Law and except by reviewers for the public press), without written permission from the publishers and authors.

Designed by Jennifer Obrey, Sun Media Group, Pawcatuck, CT
Printed and bound by Ingram Press
Library of Congress Control Number: 2016949443
ISBN # 978-0-692-10419-4

Front Cover: M.J. Brush, Ivory-billed woodpecker (Plate 2)

TABLE OF CONTENTS

PREFACE 6

I. CATESBY'S WORLD, PAST AND PRESENT 7

II. COLLECTING AND CLASSIFYING: ORGANIZING
THE NATURAL WORLD 17

III. ON THE LAND
    Magnificent magnolias 23
    Ivory-billed woodpecker 27
    Audubon and Reverend Bachman's warbler 31
    Passenger pigeon 36
    Rice 40
    Seeing the Forest and the Trees 46
        Franklinia 46
        Catalpa 53
        Carolina jessamine 56
    An Ecosystem Dependent on Fire 57
        A Forest and a Bird 57
        The Laurels 62
    Tulip tree: A long drink of water 63
    Oaks 67
    Wild Olive 70
    About Leaves 71
    Thugs and Aliens 72
        Empress tree 75
        Black locust 76
        Brazilian pepper 77
        Honeysuckle and cedar waxwing 82
        Purple loosestrife 83

IV. WHERE THE LAND MEETS THE SEA
    Salt Marsh: A dynamic equilibrium 87
    Ibises: scarlet and white 89
    Great blue heron 96
    Green heron 97
    On the beach: Some Crabs 102
    Fish crow 106
    American oystercatcher 110
    Find a spot; settle down
        Oysters 117
        Of Shells 121

V. OCEANS FORMERLY FULL OF FISH
    Fishin' for a livin' 125
        Snappers, Bass and Graysby 125
    America's Fisheries 132
    Chasing a Diminishing Resource 134
    Who will get the fishies when the boat comes in? 135
        Yellow-fin tuna 138

VI. BAHAMAS & BEYOND
    700 Islands in some of the world's clearest water 141
    Green sea turtle 142
    Catesby's Bahamas 143
        Sour orange 146
        Caribbean spiny lobster 147
        Gray triggerfish 156
    Sargasso Sea 157
    Light in the Darkness of the Deep 161
        Deep Sea Fish 166
    Homeward Bound 167

ACKNOWLEDGEMENTS 169

APPENDIX
Resources and Further Readings 170

INDEX 184

Note to readers: Numerals following Catesby quotes refer to volume and page number in the first edition of *Natural History of Carolina, Florida and The Bahama Islands*, in the Biodiversity Heritage Library (www.biodiversitylibrary.org)

PLATES

	Artist's chart of Eastern Seaboard	19
1	Magnificent magnolias (*Magnolia* spp.)	25
2	Ivory-billed woodpecker (*Campephilus principalis*)	29
3	Bachman's warbler (*Vermivora bachmanii*)	35
4	Passenger pigeon (*Ectopistes migratorius*)	39
5	Rice (*Oryza sativa*)	43
6	Franklinia (*Franklinia alatamaha*)	49
7	Carolina jessamine (*Gelsemium sempervirens*)	55
8	Long-leaf pine (*Pinus palustris*); Red-cockaded woodpecker (*Picoides borealis*)	59
9	Mountain laurel (*Kalmina latifolia*)	61
10	Tulip tree (*Liriodendron tulipifera*)	65
11	Wild olive (*Osmanthus americanus*)	69
12	Empress tree (*Paulownia tomentosa*)	79
13	Brazilian pepper-tree (*Schinus terebinthifolius*)	81
14	Morrow's honeysuckle (*Lonicera morrowii*); Cedar waxwing (*Bombycilla cedrorum*)	85
15	Salt Marsh (*Spartina*)	91
16	White ibis and scarlet ibis (*Eudocimus* spp.)	95
17	Great blue heron (*Ardea herodias*)	99
18	Green heron (*Butorides virescens*)	101
19	On the Beach: Some Atlantic Crabs	105
20	Fish crow (*Corvus ossifragus*)	109
21	American oystercatcher (*Haematopus palliatus*)	113
22	Eastern oyster, Stone crab, Shrimp spp.	119
23	Silk snapper (*Lutjanus vivanus*)	129
24	Black sea bass (*Centropristis striata*) Graysby (*Cephalopholis cruentata*)	131
25	Yellow-fin tuna (*Thunnus albacares*)	139
26	Green sea turtle (*Chelonia mydas*)	145
27	Sour orange (*Citrus x aurantium*)	149
28	Caribbean spiny lobster (*Panulirus argus*)	151
29	Gray triggerfish (*Balistes capriscus*)	155
30	Sargassum (*Sargassum natans, S. fluitans*)	159
31	Alvin & Sea Mounts	163
32	Deep Sea Fish	165

PREFACE

Mark Catesby (1683-1749), collector, artist, and author believed that "…what the Antients have described barely by words; in order to avoid such Confusions, we shall take care to Exhibit every thing drawn by the Life…" (Proposals).

His travels and the records he left became our inspiration. The watercolors that are the core of Catesby's N*atural History of Carolina, Florida, and the Bahama Islands* were his legacy and worldview. Catesby's two volumes, an amazing tale of perseverance, exploration, and art, inspired us to visit his world and see with fresh eyes how the region has changed and what has persisted. MJ, a scientific illustrator, was fascinated with Catesby's watercolors. For Alan, an ornithologist, they provided a perspective on nature, history, and the environment. Catesby represents a benchmark for the early stage of the age of exploration and discovery in eastern North America.

Catesby created one of the first realistic pictorial records of the plants and animals along the east coast of North America. The work became important in the dissemination of colonial natural history to the period's English scientific community. The images remained in relative obscurity until the late-1990s when a selection of the originals were exhibited in the United States. A catalogue of the collection and a symposium on Catesby followed. Alecto Historical Editions published a facsimile edition of the figures. Alan Feduccia's (1985) *Catesby's Birds of Colonial America* further stimulated our interests.

Natural history provides a common thread that binds together the enormous changes that have transpired in human history and the environment since Catesby's visit to North America. For this book, we selected plants and animals, some illustrated by Catesby, and others that we felt were representative of particular habitats or processes. Revisiting his world and reflecting on the times offers an opportunity to see how far our understanding has come, and how our views have evolved. We comment on how portions of this world are organized biologically, detail the roles of contemporary ecological interactions, and illustrate by both words and image many of the changes that have occurred since Catesby. Our challenge was to explore the ways plants and animals make their living in a continually shifting, changing, and often hostile world. Organizational principles such as community ecology or the ocean's food webs, physiological functions such as adhesion or photosynthesis, and human-caused changes such as introduction of alien species, habitat modification, or unsustainable exploitation of resources are employed to track transformations since Catesby's time.

We began our exploration of Catesby's world aboard our sailboat traveling to destinations he wrote of, visiting libraries and historical sites to sample history, flora, and fauna. We also worked in the collections at the Duke Marine and National Fisheries Labs in Beaufort, North Carolina. We re-visited the areas around Charleston and Savannah, and the island of Eleuthera in the Bahamas, immersing ourselves in both colonial and natural history. Beginning with Mark Catesby's published illustrations and text we compared what he and his sponsors understood of the natural world to our view from the early twenty-first century. Our hope is that by focusing on biological organization and process we might encourage stewardship of our natural world that so impressed Catesby.

I

CATESBY'S WORLD, PAST AND PRESENT

Mark Catesby (1683-1749) was most likely born at Castle Hedingham, Essex, England, the sixth of eight children. His father, John Catesby, was a landowner, involved in local government. While Catesby's early life was one of moderate privilege, there is little direct evidence regarding his early education. It is possible that he attended grammar school in nearby Sudbury. Surprisingly, there is no evidence that he showed an early interest either in art or natural history. His maternal grandfather, Nicholas Jekyll, lived nearby in Castle Hedingham and was locally well-connected. Other relatives and family friends were interested in natural history. These contacts may have shaped his interests. Catesby's early years and known family life and ties are documented in *The Curious Mister Catesby* (2015).

In 1712 Catesby made his first trip to North America accompanying his eldest sister to Virginia. He spent several weeks at the estate of William Byrd on the James River where he collected specimens of both decorative and potential crop plants. He also explored the Tidewater region of Virginia and followed the James River some distance into the Appalachian Mountains.

In 1714 Catesby sailed from Virginia to Jamaica where he continued to collect. Specimens were sent to his contacts in London and it was here that he first became aware of the work of Sir Hans Sloane. Although Catesby probably underestimated the value of his specimens and observations from this trip, he made his mark on a number of influential members of London's expanding scientific community. Botanizing was clearly a worthwhile endeavor. Catesby was recognized as a careful collector, interested in plants of potential horticultural importance and also those decorative plants that might be cultivated in England. The long-term consequence of this trip was his introduction to the emerging London intelligentsia, many of whom were botanists who used to met informally at London's Temple Coffee House, then a focal point for scientific discussions and the promoting of botanical knowledge (Riley, 2006).

Catesby's second voyage came about with the support of an influential group of sponsors who shared an interest in the natural world. His assignment in the New World was "to observe the rarities," collect useful plants, and to search out "natural curiosities." Samuel Dale, an apothecary interested in medicinal herbs, knew of Catesby's earlier Virginia trip. Dale saw to it that specimens received from Catesby were brought to the attention of William Sherard, then considered England's leading botanist. Sherard, in turn, acted as a conduit to the London establishment. Further support for Catesby was garnered from the merchant Peter Collinson (1694-1768), a Quaker and active member in the Temple Coffee House group who worked hard to collect plants and seeds. Later in the century

Collinson provided Catesby an interest-free loan to finish his book. Along with Catesby, he also sponsored John Bartram (1739-1823) of Philadelphia, who botanized, observed everything, and eventually became an early force in botany in America.

Some of these men were Fellows of The Royal Society of London which had been granted its charter by Charles II in 1662. They formed an enthusiastic group of physicians and natural philosophers who since 1687 had encouraged colonists to contribute interesting specimens. Sherard regularly attended Royal Society meetings in London. Sir Hans Sloane (1660-1753) was also actively involved. Sloane had spent three years in Jamaica as personal physician to the Duke of Albemarle and collected about 800 of the 2500 plant species of the island. Subsequently, these specimens became the nucleus of an enormous personal collection that eventually provided the basis for the British Museum. Another apothecary, James Petiver (1665-1718) had supported the earlier work of John Lawson (1674 to 1711), the Surveyor General of North Carolina.

Lawson had begun a natural history survey of the Carolinas, which was left incomplete when the Tuscarora Indians killed him. Petiver actively supported the natural sciences and in 1690 he wrote one of the first instruction sheets in English for collectors. Unfortunately, Petiver died shortly after Lawson and the plans to complete the project fell to Sherard. The interests of Sherard, James Brydges (first Duke of Chandos) and others reflected their influence and prominence as collectors of exotic plants or patrons of the arts. Colonel Francis Nicholson, the newly appointed Governor of South Carolina, offered financial and logistical support that facilitated Catesby's trip.

Botanical studies of the time often focused on the medicinal uses of different plants, hence the large number of physicians and apothecaries among Catesby's sponsors. European settlers in the tropics learned that natives used the alkaloid quinine derived from the plant *Cinchona officinalis* to combat malaria. *Caesalpinia pulcherrima*, a plant native in Southeast Asia, but now pan-tropical, was also available for use by physicians. Natives used this plant as an abortifacient, but political considerations mitigated against its availability and acceptance in Europe. These instances were each a part of the emerging changes in scientific observation and the practice of medicine. The colonies were a source of both garden plants and herbal medicines.

Catesby's abilities and experience fitted his patrons' needs and expectations. Unfortunately for today's historians, Catesby wrote only brief passages to accompany his watercolor plates. His emphasis was on simple descriptions with little information on distribution or reproduction as might be found in a modern field guide; however, Catesby's narrative text gave considerable attention to the agricultural and economic potential of the areas he explored. While he apparently kept no log, a number of his letters to his sponsors remain. His close contemporary John Lawson (1709) and later William Bartram (1791) wrote more extensively; both used a diary or log-like format to describe and report their trips. But their images were few and their goals were somewhat different. Without doubt it was Catesby's ability with watercolors that made his contribution so valuable and unique. He produced some of the first pictorial representations of North American plants and animals by a European which would not be matched in scope until John James Audubon (1785-1851) appeared on the scene 100 years later.

Catesby was a competent observer with a somewhat broader worldview than his predecessors. For example, Catesby wrote of the "Plat Palmeto":

"This Palm grows not only between the Tropicks, but is found further North than any other. In *Bermudas* its Leaves were [f]ormerly manufactured, and made into Hats, Bonnets, &c. and of the Berries were made Buttons. This is the slowest grower of all other Trees, if Credit may be given to the generality of the Inhabitants of *Bermudas*, many of the principal of whom affirm'd to me, that with their nicest Observations, they could not perceive them to grow an inch in height, nor even to make the least Progress in fifty Years, yet in the Year 1714, I observ'd all these Islands abounding with infinite Numbers of them of all Sizes: This kind of Palm grows also on all the Maritime Parts of *Florida* and *South Carolina*, whose Northern Limits being in the Latitude of 34 North, is also the farthest North, that these Palms grow to their usual Stature, which is about 40 Feet high, yet they continue to grow in an humble Manner, as far North as *New England*, gradually diminishing in Size, as they approach the North, being in *Virginia* not above four Feet high, with their Leaves only growing from the Earth, without a Trunk, yet producing Branches of Berries, like those of the Trees: In *New England*, they grow much lower, their Leaves spreading on the Ground: This remarkable Difference in the same Plant has been the Cause of their being thought different Species, tho' I think they are both the same, and that the smallness of the Northern ones, is occasioned by their growing out of their proper Climate, which is hot, into a much cooler one, where the Heat of the Sun is insufficient to raise them to Trees" (I-xli).

This perceptive description of variation within a species induced by an environmental influence brought Catesby to the brink of a critical insight; however, it passes without comment. For whatever reason, populations of this plant show variation associated with distribution. What is implicit in variation of this type, which is indicative of potential change, was to be explored over 140 years later in Charles Darwin's theory of the origin of species. Variation is significant, as it is the material on which selection works. It is the engine for change over time. In the time of Catesby, species (as well as geological features) were believed to be fixed. The world, Nature itself, was complete and perfect at creation; subsequent change was due solely to deterioration from the created goodness. The widely held belief was that humans were at the center of the world and that all else had been created for man's use or enjoyment. Galileo had suggested earlier that the earth was not at the center of the solar system. A purely naturalistic origin of the earth, the solar system, and the stars did not emerge until the early nineteenth century, followed a half-century later by an understanding of the origin of species. This was a natural world with neither a God nor man at the center, or, indeed, with no center to be inhabited by God, man, or any other. Such thinking would challenge directly the traditional concepts of purpose and the meaning of life as understood in Catesby's time.

Catesby noted that: "In *America* are very few *European* Land Birds, but of the Water Kinds there are many, if not most of those found in *Europe*, besides the great Variety of Species peculiar to those Parts of the World" (I-xxxv). This is clearly an observation with biogeographic acumen. Animals can be different in different parts of the world, not the uniform distribution posited by creation myths. Catesby also was comfortable with the thought that seabirds moved freely across the oceans. They appeared far out at sea and minded no apparent boundaries; hence the seabirds in the New World resembled those in Europe. Speculating about the relation between the two, Catesby assumed the priority of the Old World over the New:

> "Admitting the World to have been universally replenished with all Animals from *Noah*'s Ark after the general Deluge, and that those *European* Birds which are in *America* found their Way thither at first from the old World…In the Island of *Bermudas* it frequently happens that great Flights of Water Fowl are blown from the Continent of *America* by strong North West Winds, on that island [Bermuda]" (I-xxxv).

While some bird species may have spread from Old to New, the presence of entirely different species seemed to contradict assumptions about the unitary nature of Creation. Further, by these times naturalists were aware of the existence of fossils but not necessarily their significance in understanding evolution. Because species had been created by God, extinction was not considered an option; if species were lost, it implied something amiss in the Great Design. But if these species were annihilated in the biblical flood, as part of God's intended punishment, then there was a reason for their demise (and subsequent fossilization). Multiple creations were acceptable to explain the repopulating of the world by the animals on the Ark. Catesby had witnessed mastodon teeth dug from the ground near Stono, South Carolina. However, he seemed to be working toward a different explanation when he wrote:

> "…the Cause of Disparity in Number of the Land, and Water Kinds, will evidently appear by considering their different Structure, and Manner of Feeding, which enables the Water Fowl to perform a long Voyage, with more Facility than those of the Land" (I-xxxv).

In other words, very much like the geographic differences in Plat Palmetto, there are other forces at work and somehow animal and plants adjusted through time to environmental differences. But that may be getting ahead of the story as Catesby's text makes no mention of cause and effect. He is dealing with something he finds notable, but cannot completely rationalize.

In considering waterfowl, Catesby invokes strong winds to explain their distribution among different islands. But what of the occurrence of land birds? The geography and the avifauna of the high Arctic were unknown in the early eighteenth century. Catesby reasoned:

> "Tho' the nearness or joining of the two Continents be not known we may reasonably conclude it to be within or very near the Arctic Circle, the Coasts of the rest of the Earth being well known; so that those few *European* Land Birds that are in *America*, must have passed thither from a very frigid Part of the old World, and though these Birds inhabit the more temperate parts of *Europe*, they may also inhabit the very northern parts, and by a firmer Texture of Body, may be by Nature better enabled to endure extream Cold than Sparrows, Finches. and other *English* Birds, which are with us fifty to one more numerous, but are not found in *America*" (I-xxxv).

Catesby was speculating, of course, about the presence of these birds in North America, and he was uneasy with this explanation. He continues:

> "Tho' these Reasons occur to me, I am not fully satisfied, nor do I conclude

that by this Method they passed from one Continent to the other, the Climate and their Inability of performing a long Flight may reasonably be objected. To account therefore for this extraordinary Circumstance there seems to remain but one more Reason for their being found on both Continents, which is in the nearness of the two Parts of the Earth to each other heretofore, where now flows the vast *Atlantick Ocean*" (I-xxxv).

Moreover, Catesby goes on to remark that some of these species, for example kinglets, are very small and long distant flights are unlikely. The suggestion was that, sometime in the past, perhaps in post-flood times, the as-yet-unknown Arctic was a single land mass. This, of course, anticipates the question regarding the permanence of the continents, which wasn't thoroughly understood until the late twentieth century.

Yet Catesby still had made the undeniable observation that many birds in the Carolinas and Europe are "birds of passage." That is, these species were migratory. He opines that this capacity can explain long distance moves, even potential transcontinental passages. He admits that "…no ocular Testimonies have been produced…"(1-xxxv). But those species in the Carolinas may indeed leave and "…that the Place to which they retire is probably in the same Latitude of the southern Hemisphere, or where they may enjoy the like Temperature of Air, as in the Country from whence they came…"(1-xxxvi). These ideas were presented to the Royal Society and were considered a valuable contribution to understanding both the migration and the distribution of birds in the New World. How far have we progressed? Indeed, at the start of the twenty first century we have considerable "…ocular testimonies…" to the occurrence of migration on many continents, but a complete mechanistic explanation of migration is still lacking. Science has still not illuminated all the niceties of, for instance, how migrants compartmentalize the energy to support long distance flight, how they mark their position, or even how they migrate while sleep-deprived.

Catesby's bird list totaled 109 species. There was some overlap with the earlier list generated by John Lawson. Lawson had produced only a list, no illustrations, and obtained no specimens. Catesby provided both illustrations and information. Catesby provided the essential documentation of what he observed. There was, at the time, no formal way that organisms were classified. The tools to infer relationships and relatedness simply did not exist. Of course, without the theory of evolution there was no theoretical need to know; however, knowledge of the natural world was increasing to the point where it was becoming necessary to at least categorize in a useful manner. Being able to identify plants and animals consistently was essential to commerce and the interactions among interested parties. Increased information leads also to reorganization and management of the emerging appreciation of the varieties of plants and animals of the living world. Once again, Catesby is on the cusp of crucial new understandings of the natural world, but not quite there yet. In his day no one had the vision to consider that there may be a rational use in systematizing the knowledge about the form of organisms and their parts.

At the time of Catesby's birth, species were considered immutable. Received knowledge held that each type of organism had been set at the Creation, designed for its particular role, and then simply existed unchanged. This scheme of life is traceable to Aristotle (384-322 BC), and was part of a divine plan that was related to the organism's moral value. The scheme included a ladder of life that scaled animals from lowest to highest, with man as the pinnacle. Time since the creation was not long and only a relatively few generations separated the late seventeenth century from the Garden of Eden. There was little or no attempt to understand function, much less consider genetic relatedness or the origins of species. However, in this age of exploration, naturalists were becoming aware that the world of creation was far richer than had previously been imagined. To comprehend the richness of the natural world, naturalists had begun to use structural traits to identify plants. From this, a crude classification system developed. One, it would turn out, that was based on the features of the organism rather than a Divine plan. But that gets ahead of the story.

Perspective on the grand design of nature was shifting by the late 1600s and Mark Catesby would be influenced by the new currents of thought. John Ray (1627-1705) was an established influential naturalist and an active member of the local community of lawyers, antiquarians, and horticulturists. Educated at Cambridge in mathematics and science, Ray subsequently trained as a clergyman, not as a physician as were many of his contemporaries. He was a leader in studies of natural philosophy, published prolifically, and by age 40 was a member of the Royal Society. His first book, based on the efforts of students at Trinity College, Cambridge, catalogued 626 local plants. This opened a window on a world more diverse than had been imagined previously. In time, Ray edited *Ornithology* (1676), a catalogue of the known types of birds prepared by his friend Francis Willughby, and published several more books on plants, and two on natural theology. Ray became interested in natural history as a consequence of a search for a system to explain the diversity of Creation.

In *The Wisdom of God Manifested in the Works of Creation* (1692), Ray argued that a natural theology could be based on studies of God's creation, which is the natural world. Ray held that pondering nature honored God's infinite wisdom. He came to consider as a part of the plan the adaptations of animals and plants and the relationships of form to function. These ideas were new, but not heretical, and part of the growing dissonance between science and the Church. Ray believed both in God's creation and the value of classifying living things in an attempt to understand nature's Divine magnificence. These observations were loaded with important implications and would lead eventually to questions regarding physiology, function and behavior.

John Ray was on the forefront of those who struggled to recognize the existence of diversity in the organic world. One aspect of the challenge they faced was the problem of how to classify and identify organisms. Ray's approach to classification was essentially pragmatic. He used structure, life history, and even ecology as traits. Each kind, or species, consisted of similar individuals. Ray also realized that several kinds could share a number of common attributes and might form a group that resembled one another. Ray believed that these were natural groups, which should be reflected in classification. He came to think that classification must be based on these observable attributes, rather than, say, usefulness to man or a symbolic or moral value. To reconcile these concepts with his appreciation of the grand design it was understood that classification helped describe life's diversity, but did not explain creation or an animal's or plant's place in God's purpose.

Ultimately, Ray's influence on the modern classification of animals and plants was limited. This was due to the fact that the natural philosophy of the time, still conforming to the Aristotelian view, argued forcefully for final causes in design. Nevertheless, Ray and his contemporaries were taking decisive steps in a new direction by basing their classifications not on classical Aristotelian logic, but by relying on their own observations based on larger collections and careful analysis of specimens. From this foundation rational classification eventually followed.

The task eventually fell to Carl Linnaeus (1707-1778) of Sweden, whose book *Species Plantarum* (1753) introduced a hierarchical system of classification based on flower parts (flowers are often referred to in the vernacular as the sweet smelling sex organs), as well as binomial nomenclature. Previously, plant names had been simply descriptive and long. Linnaeus's system included a concept of species grouped into increasingly inclusive units. The system was a practical tool and useful for identification and information retrieval. His species, however, remained fixed and divinely created. Modern botanical nomenclature began with Linnaeus's *Species Plantarum*. In the 10th edition of his *Systema Naturae* (1758), Linnaeus included animals and a similar system for their nomenclature. Catesby's *Natural History*, completed earlier, had no formal standing for modern names. However, his plates provided Linnaeus the 'specimens' for 81 North American and Bahamian bird species.

The archaic, fixed concept of species did not change until the mid-nineteenth century with the publication of *On the Origin of Species by Means of Natural Selection, or Preservation of Favored Races in the Struggle for Life* (1859) by Charles Darwin. He held that species were not immutable, nor were they divinely created. In the mid-twentieth century, systematics (the study of evolutionary relationships of organisms) and taxonomy (the naming and classification of taxa) underwent another major retooling. Genetic and molecular data and biogeography are now used to reconstruct the actual events that generated the patterns of biodiversity. Strictly defined basal and derived characters are used in the construction of clades, those groups of organisms that consist of the single common ancestor and all the descendants of that ancestor. The new systematics is based on phylogenetic relationships, rather than perceived similarities, which can be misleading. This approach has led to a reappraisal of the concept of *species*, a debate with consequences for conservation and understanding biodiversity.

Paradoxically, Catesby has no species of bird named after him. The best known animal bearing his name is *Lithobates catesbeiana*, the bullfrog, so popular in biology class dissections. *Catesbya pesudomuraena*, an eel found in the Bahamas, is another animal named for him. Plants named in his honor include the native pine lily, *Lilium catesbaei*, and *Catesbaea*, a large genus of flowering plant that occurs in the Florida Keys, Bahamas, and the West Indies.

After his death, the watercolors, plates, and the remaining copies proved to be of value to his widow. Fortunately for us, the sketches and watercolors have been stored in Windsor Castle since then. King George III bought all the material from a London dealer in 1768 for £120.

When the vessel carrying Mark Catesby approached Charleston in mid-April of 1722, strong winds blew off shore. Typical of the barques of the day, the ship was three-masted with the foremast and mainmast square-rigged, and the aft (mizzen) mast rigged fore-and-aft, but it did not sail well to windward. Sea conditions dictated a delay outside the harbor. The master and crew found themselves back in the currents of the Gulf Stream with the high seas that are generated under these blustery conditions. The Captain was forced to wait two weeks for more propitious conditions to beat into the harbor. It was under these conditions that Catesby collected his first bird, a ruddy turnstone (*Arenaria interpres*).

Charleston was a natural port of entry for the Carolinas. In 1722 it was a center for commerce with an established community of successful landholders, tradesmen, and farmers. At the time, the well-protected harbor was shallow and given to shoaling. The Ashley and Cooper Rivers flow into the harbor and provide water access inland; they are also the source of the silting. The inlet from the sea is tidal and river currents can be strong, especially during the rainy season. Arriving ships anchored in the river channels and unloaded goods, passengers, and crew by ship's launch. Charleston Harbor today is still lively with the skyline dominated by cranes, bridges, and the silhouettes of monstrous, world-girdling container ships.

From the water, it is easy to imagine Charleston in 1722 even though only a few buildings of that era survive, for example the Powder Magazine at 75 Cumberland, the Pink House Tavern at 17 Cumberland, and 106 Broad Street. Queen Anne's War was fought on nearby Sullivan's Island in 1706, and status as a Royal colony had been granted only in 1721. The main export goods were tobacco and forest products such as pitch, tar and the wood that provided planks and spars for an active shipbuilding industry. Eventually rice and indigo were added.

Governor Nicholson met Catesby at quayside. The connection proved valuable. Nicholson introduced Catesby to the local gentry who became hosts and facilitated travel inland and up and down the coast. Catesby quickly began to accumulate specimens and traveled inland along with nearby marshes and coastal areas. Local fishermen most likely provided fish specimens. The marshes of the Cove near Fort Moultrie and Sullivan's Island, Kiawah, Johns and James Islands to the south were productive, accessible collecting locations for gulls, terns, cormorants, herons and ibis. Catesby noted that on Sullivan's Island:

> "…which is on the North Side of the Entrance of *Charles Town* Harbour, the Sea on the West Side has so incroached (tho' most defended, it being on the contrary Side to the Ocean) that, it has gained in three Years Time, a Quarter of a Mile laying prostrate, and swallowing up vast Pine and Palmetto-Trees. By such a Progress, with the Assistance of a few Hurricanes, it probably, in some few Years may wash away the whole Island, which is about six Miles in Circumference." (I-iii).

Fortunately, this hasn't happened, but the comment demonstrates his understanding of natural forces causing erosion.

> "At about half a Mile back from the Sand-banks before-mentioned, the Soyl begins to mend gradually, producing Bays, and other Shrubs; yet 'till at the Distance of some Miles, it is very sandy and unfit for Tillage, lying in small-Hills, which appear as if they had been formerly some of those Sand-Hills formed by the Sea, tho' now some Miles from it" (I-iii).

If you factor out the human construction, the islands, creeks, and tidal marshes of the Cape Romain National Wildlife Refuge probably appear now much as they did in the early 1700s. There is a sense of place on the marshes nearby. On a sunny day the view is vast. At night, only the calls of the local rails break the silence.

We arrived in Charleston the first time 278 years after Catesby on our boat under sail. We passed through the Sullivan Island Narrows which, despite Catesby's prediction, still exist. The Narrows are an integral part of the Intracoastal Waterway (ICW). The islands above Charleston are flat and low, carved by a series of creeks and inlets. A number of shallow tidal creeks wander through the flats. The Ben Sawyer Swing Bridge, obviously not present in 1720, connects Sullivan Island to the mainland. Fort Moultrie is located on the southwest corner of the island, facing the sea. It was the current through Breach Inlet that, in 1776, defeated the British as they tried to wade from Isle of Palms to attack the fort. Fort Moultrie has never been taken and, even today, signs warn swimmers against attempting the currents.

The Intracoastal Waterway (ICW) is a dug channel here. The ICW was authorized by Congress in 1919 and runs 3000 miles along the eastern and Gulf coasts of the United States. A man-made feature of the coastal environment, it is maintained at a minimum width of 80 ft. and a charted depth of rarely more than 9 ft., frequently less. Beyond the bridge the most prominent feature is Charleston Light. At 183 ft. tall it is visible at great distances and defines the north side of the harbor entrance. Cummings Point and Fort Sumter, with a garrison-sized Old Glory, mark the south side. For several miles seaward the approach to the harbor mouth is generally shallow which reflects the gentle slope of the continental shelf. Submerged jetties extend seaward along with the navigational aids that indicate the man-made channel. The 45 ft. deep channel begins at a sea buoy and a range marker guides vessels into Charleston Harbor.

Charleston today is a modern city with a bustling harbor. Recreational boats still anchor off the seawalls in the Ashley River just as in the early 1700s. Accordingly, it is necessary to mind the currents and the winds, and make sure the anchor is well set. How much the harbor and adjacent coastal habitat resembles that of Catesby's day is another matter. Surely, cutting channels and dredging have affected the patterns of current flow. Clearing the flooded forests upstream and changes in agricultural practices have influenced the patterns of erosion and sedimentation. Regardless of the changes, the deck of a small sailboat offers an unprecedented view.

The Carolinas, like the rest of North America, have changed extensively since Catesby's time. The unrelenting increase in human numbers and activities has been the primary force driving the change. As early settlers attempted to adapt a life style from the Old World to that of the New, the face of the land was changed. Little of the original forest remains. Every major river in the Carolinas is now dammed. Consequently, exploratory trips based on river travel such as those made by Lawson (1709), Catesby (early 1720s), and the Bartrams (1773 thru 1778) are no longer possible.

The intended and unintended consequences of human activity subsequent to Catesby's *Natural History* were as unpredictable as they were monumental and widespread and these concern us. It is highly unlikely that the accumulated damage can be repaired, or pristine conditions restored; nor would we necessarily want to return to the living conditions of 300 years ago.

Nevertheless, species continue to go extinct at an ever-increasing rate. The atmosphere is warming at historically unprecedented rates. The far-reaching consequences of human-caused climate change are melting glaciers, changing sea levels and ocean chemistry, shifting ocean currents, disrupting the atmospheric ozone layer, and disturbing entire ecosystems. The oceans are massively overfished and the near-shore environment altered. Native grasslands have vanished. Fertilizer used to increase the biomass yield of cereals for human consumption and silage for livestock has polluted waterways. Coastal marshes have been drained, filled and developed, interfering with the breeding of numerous marine organisms and setting off cycles of coastal erosion. The changes in water aquifers and surface water courses eventually change natural habitats and cycles, and can combine with drought to cause further destruction by fire, wasteful consumption, and contamination. Native forests are cleared, or fragmented, and the land converted to human use. As a consequence, plant diversity has shifted, use-patterns by animals interrupted, and food webs truncated. Large land predators (wolves), grazers (bison, caribou, elk) and fur-bearing species (seal, beaver) have been relentlessly exploited. Similarly, in the ocean, the top predators both fish (tuna, swordfish and sharks) and mammals (whales and seals), have been dangerously over-harvested by both commercial and recreational fishers.

We feel that the ecological consequences of these activities are best understood in historical perspective and that the innate beauty of the plants and animals can expand our understanding and enhance our appreciation of the natural world. Of course, none of this is limited to the Carolinas, but Catesby's world provides a benchmark. The original bounty is gone, as are the pristine conditions that accommodated it.

II

COLLECTING AND CLASSIFYING: ORGANIZING THE NATURAL WORLD

A Brief History

The narrator of Genesis reported that Noah was told to build an ark of gopher wood in anticipation of an oncoming deluge and flood. Noah was just over 600 years old and besides boarding his wife, three sons and their wives, he was to accommodate at least two of every beast (both clean and not), fowl of the air, and all things "that creepeth upon the earth." Bringing each by two's insured one of each sex as well. Presumably, the fresh-water and marine animals were not worried about the impending flood, being successfully adapted to the aquatic environment. Accordingly, Noah became one of the first to categorize the *kinds* of living things, although the record doesn't show that he devoted a lot of thought to the problem. In contemporary terms he executed an inventory of the local biodiversity.

Placing things in categories seems to be a universal human trait. Even without the threat of immediate inundation, native peoples worldwide have been found to be remarkably adroit at this task and are capable of accurately recognizing and naming most, if not all, of the plants and animals in their environment. Most employ a criterion of similarity, which works well for defining membership in the particular category. This capacity for sorting and naming is of great survival value. It is essential in communicating about plants and animals.

The effort to inventory the world's diversity, especially plants, was one of the intellectual highlights of early eighteenth century Europe. The French, Scandinavians, and most especially the British, wanted to impose an order on plants and animals, perhaps not unlike the order Newton envisioned with the laws of universal gravity and motion only 30 years earlier. Myriad new plants and animals were being discovered in the remote parts of the world. Catesby was but one such collector in North America. Classification followed acquisition, and the prospect of organizing useful categories faced ever-mounting challenges with the rapidly expanding base of plants and animals to be classified. As was mentioned earlier, in 1689, John Ray, a contemporary of Newton and Edmond Halley (1656-1742), proposed a classification scheme for plants based on similarities of the root, flower, calyx, seed and seedpod. Ray had a passion to organize, and understood instinctively that certain plant and animal types differed from all others. These patterns and sub-patterns were useful in identification and Ray called the basic unit (the Biblical *kind*) *species*. Each had its own name by which it would be distinguished from similar groups. The Frenchman Joseph Pitton de Tournefort (1656-1708) working along similar lines, named several plant genera, a term he used for groups of similar species, thereby adding another element to a basic hierarchal system that would be a progenitor of our own. A species concept was now in hand and simple categorizing had begun to expand towards a classification. Nevertheless, innovative as they were, these attempts to classify plants were hardly adequate for the task ahead.

While these assemblages are woven through the botanical literature of the times, it was Carl Linnaeus who revolutionized the nomenclature of both plants and animals. So influential was Linnaeus that some considered him the premier biologist of his time; he was certainly one of the most conspicuous. His taxonomy was promulgated on the idea of species espoused by Ray and the notion of a hierarchy of life. His *Species Plantarum* (1753) was based on the fact that plants, like animals, were sexual and the stamens and pistil of the flower provided the key to his system. The use of the reproductive parts for classification brought an artificial semblance of order. While sexuality was embarrassing to some at the time, the morphology of reproductive organs provided a logical framework to the chaos of the existing systems. Linnaeus's system provided a hierarchy of increasingly inclusive categories – species, genera, family, order etc., that were defined by similarity within each division.

A feature of Linnaeus's vast collecting activities was that an example of the organism (sometimes even a drawing or painting) existed that corresponded to the new binomial name established for each species. The name resides in a single concrete individual—the preserved specimen—that was designated by the person who first published on the newly described species. Publication establishes priority that goes to the author, not the person who finds or collects the specimen. This individual specimen is designated the 'type'. It maps the formal name to a particular individual specimen, ideally for all time. The name, in Latin, conforms to the Linnaean binomial model. The name compresses the many (the population) into one (the specimen), which renders the abstract into the concrete.

True, there have been and still are, confounding issues with the system, but it has survived and provided stability in a field that has absorbed dynamically expanding data. It is important to realize that the type is not now considered to be a prototype or archetype of the species. That is, that the object in the 'real' world was represented in an ideal form that existed in a Divine mind of some sort: the Platonic doctrine of *eidos*. Linnaeus may have believed that the essence defined the species. But his system has endured because it coincidentally fits the evolutionary concept that lineages are maps of descent that branch and turn through time. The philosophical core of the Linnaean system crumbled over time in light of subsequent work that showed species were not immutable, that genera came and went over time, and that the geographic distribution of species did not follow a rational pattern.

Linnaeus's second contribution, also introduced in *Species Plantarum* (1753) and extended in the tenth edition of *Systema Naturae* (1758), was the formalized use of a binominal system of nomenclature. The long, often ungainly descriptive names for plants and animals used at the time (for example, *Quercus Anpotius Ilex Marilandica folio longo*) the willow oak, were replaced by a generic name (*Quercus*) and a specific epithet (*phellos*). The species itself, the basic taxon, was considered static and unchanged since its first appearance on earth. The binomial system, simple and based on direct observation, offered an easily understood map to nature. Subsequently, priority of publication is used to assure the singularity of the specimen.

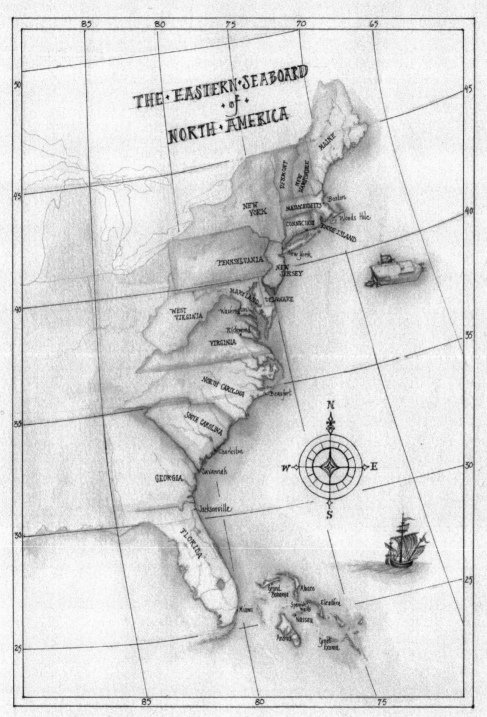

Linnaeus, the stubborn Swedish provincial, was a conservative, poorly dressed, argumentative professor, and a shameless self-promoter. He provided anyone interested in the collection and study of plants with an instructional handbook and equipment lists. His "disciples"--students and travelers--botanized all over the world, and contributed the specimens that, supplemented by Linnaeus's own Herculean activities, built a large museum collection at Uppsala University. There was also an extensive botanic garden with many plants associated with him that still exists today. With zeal and insight, he named plant species for their discoverers, various friends, and numerous associates.

Other naturalists, including the English horticulturists, medical herbalists, and various acquisitive landowners, had sophisticated collections and gardens in the early 1700s. It was heady to grow an eponymous species in one's garden. Linnaeus's system of classification was embraced instantly by naturalists throughout Western Europe. Later in the century the intrepid Captain James Cook returned from his first voyage with about 1300 plants previously unknown in Europe, collected by Joseph Banks and Daniel Solander. Linnaeus's system was welcomed by Banks (1743-1820), who later became the long-term president of the Royal Society. Banks, in turn, had great influence on science and society in Britain. Both King George II and King George III took an interest in natural history, and supported these collecting efforts. The Dowager Princess of Wales, Princess Augusta, mother of George III, was involved with the creation of Kew Gardens.

Two of Linnaeus's works, *Systema Vegetabilium* and *Genera Plantarum,* were translated into English by Erasmus Darwin (1731-1802). In translation, the titillating aspects of Linnaeus's plant classification, with its implied sexuality, was not lost on the public. Darwin, a physician with broad interests in natural philosophy, also wrote the epic poems, *The Botanic Garden* and *Zoonomia*. In *Zoonomia* he developed a controversial evolutionary theory, vaguely related to Ovid's classic *Metamorphsis*. Ovid proposed that all living things had a common origin from a single living "filament," that plants, at least, struggle for light and water, and that species may change (metamorphose) over time. It is significant, in part, because during these years European intellectuals were exploring the idea that life was not fixed at the moment of creation, but had changed through time. Erasmus Darwin, one of England's intellectual leaders, was grandfather to Charles Darwin, who in the mid-nineteenth century published his theory of the origin of species based on natural selection.

A rigorous Linnaean approach to the classification of plants was not undertaken in England until about 1772 with the appearance of the two volume *Botanical Arrangement* by William Withering (1741-1799). It differed from existing herbals and the horticultural books of the day as it launched a rigorous Linnaean approach to the classification of plants. Medical usage of plants was deliberately omitted, as Withering considered such traditions based on misplaced faith and superstition. Ironically, a decade later he published his *Account of the Foxglove* (1785), which included both the plant's medical uses as well as its harmful side effects. Perhaps of more significance was that the work was based on human experimentation with the plant, something previously unknown to curious physicians.

When it came to birds, the Linnaean system was not as successful as it had been with plants. Traditional avian classification was based on the influential observations of Aristotle who divided all birds into five groups. The basis for this was vague in its details but included the belief that everything was created to fulfill a purpose, and that the world was organized in a hierarchy leading from the lowest to the highest. The Aristotelian system held sway until 1676 and the appearance of *Ornithology* by Francis Willughby. *Ornithology* was an attempt to record everything known about birds and favored a system based on form rather than function. Willughby died prior to its completion, and it was eventually finished and published by John Ray. Linnaeus had collected birds himself and received specimens from many others. He also had access to the collection of Olaus Rudbeck, Jr. (1660-1740), considered the first Swedish ornithologist. Parenthetically, Linnaeus named the genus *Rudbeckia*, coneflowers, after Olaus and his father, also a noted scientist. At the University of Uppsala we had the opportunity to study the watercolors drawn by Rudbeck's daughters.

Unfortunately, Rudbeck's bird collection was destroyed in a fire, leaving only a series of colored drawings. Linnaeus lacked access to a world-class bird collection and was thus denied the inherent power of broad, comparative studies. Consequently, his classification was imperfect. Linnaeus's treatment of birds, however inadequate, ultimately became the basis for all zoological classification and nomenclature.

The leading work on birds at the time was *Ornithologie* (1760) by Mathurin Jacques Brisson (1723-1806). Brisson was curator of the Réaumur bird collection, one of the two largest in Europe. René de Réaumur (1683-1757), had been the first in France to have an extensive natural history collection. He collected widely and kept detailed records, including information on habitat and behavior. Réaumur often collected nests and multiple specimens, which was extremely unusual for the time, but eventually inordinately valuable. As curator, Brisson created a catalogue of the collection not to be surpassed until the appearance in 1874-1895 of the 27 volumes of the *Catalogue of the Birds in the British Museum*. Brisson described the birds in great detail and defined groups with much more precision than had others working in this field. By employing more orders and genera he was able to assess relationships among specimens more accurately. Brisson was so meticulous a worker that when the International Committee of Zoological Nomenclature adopted the twelfth edition of Linnaeus as the starting point for nomenclature, they retained Brisson's genera. Although he didn't consistently use the binomial system, the work was far superior to anything previous.

When Réaumur died, King Louis XV, under the influence of the Comte de Buffon (1707-1788), ordered the collection sent to the Cabinet du Roi, against the stipulations of Réaumur's will. It turned out that Buffon, Director of the Jardin du Roi in Paris, was one of Linnaeus's severest critics. The hierarchy of nature that Linnaeus recognized reflected a God-given harmony. The essential unfolding of life's history was a predetermined sequence that led to man. In Buffon's view, all classification systems were artificially contrived. The natural world, he posited, for its entire existence (all seven epochs according to the Comte) had undergone constant change perhaps ever since the Creation. Second, Linnaeus had introduced a concept of human races based on preconceived biblical notions and not on direct observations. Buffon held that human differences were based on climatic influences, essentially a case of evidence-based arguments versus Biblical revelation. Nevertheless, Buffon, who was responsible for the transfer of the collections, apparently hated Réaumur and forbade Brisson access to the now even larger collection. Eventually, Brisson left ornithology, but realized that an accurate classification of birds would involve a collection of many more specimens, including the males and females of every species -- Noah's two of every kind. Buffon and Linnaeus manifestly held very different opinions on the nature of species and taxonomy and their differences were not quickly resolved.

In 1779 Johann Friedrich Blumenbach (1752-1840) called for classification to be based on the entire bird anatomy, not just a single character or function as Aristotle and others had done previously.

Subsequent workers modified the Linnaean classification by employing traits such as foot anatomy, bill structure, or the pattern of the sternal notch. Single traits such as these are highly modifiable in nature through adaptation and alone not sufficient for accurate classifications. It took Peter Simon Pallas (1741-1811) to carry out a synthesis of the works of Linnaeus and Buffon. He embraced Linnaeus's general systematics and the binomial system. From Buffon he took the idea that species are mutable, especially the concept that climatic conditions can affect morphology. It was Pallas who first called attention to geographic variation of animals, an idea that Charles Darwin incorporated into his theory. Taxonomists eventually came to appreciate that all individuals within a species were not identical, a further key to understanding evolution.

Linnaeus's classification system was not based on much more than overall resemblance. The individuals that resembled one another most closely were included in the same species. Similar species were then sorted into genera, genera grouped into families and so on. The result was a hierarchy that produced a classification, but the emphasis was primarily on naming each taxon and not necessarily in reflecting the mutability biologists were just beginning to recognize. A categorizing of this static nature, however practical, does not necessarily lead to understanding. The system did not reflect evolutionary relatedness or phylogeny. Darwin accepted this scheme, but later made an additional, extremely important contribution when he postulated that classification must reflect genealogy. His argument was based on propinquity of descent. The characters or traits that produced resemblance within a species were in fact inherited from a common ancestor. They were homologous. It followed that if patterns of variation based on kinship were explained by descent with modification, the resultant classification would be strictly genealogical. Implicit in this thinking was that all life descended from a common ancestor and that taxa at each level were related through common descent. That taxa would include each other's closest relatives was axiomatic, also that the lineages changed over time. Further, Darwin specified natural selection working on natural variation as the force driving change, but did not propose a method for constructing classifications that contained monophyletic groups. This problem and a universally accepted species definition are still argued today.

III

ON THE LAND

Magnificent Magnolias

Illustrations of at least three species in the genus *Magnolia* appear in the Catesby volumes. The first volume shows the small sweetbay (*M. virginiana*), which provides the habitat for the blue grosbeak (*Guiraca caerulea*). In the first plate, the sweetbay includes the blossom, leaf, and seed pod; the second plate lacks the seed pod. The seed pod is important as it figures in classification. In this illustration the pod seems to more closely resemble that of the Southern magnolia than the sweetbay. Catesby reported seeing the "Blew Gross-beak" only in the Carolinas. In fact, the bird, at least now, is more widespread and breeds as far north as Maryland. It was never abundant within its entire range and Catesby may simply have missed it. In a second, later plate, the sweetbay provides background for the males of the painted bunting (*Passerina ciris*) and the indigo bunting, (*P. cyanea*) in breeding plumage.

The much larger southern magnolia (*M. grandiflora*) or bull bay is illustrated in several of Catesby's publications. In *Natural History* it is labeled "*Magnolia flora alba*" and given no common name. The original illustrations on vellum are attributed to Georg Ehert, but their history is confusing (Nelson, 2014). A tree grown in the garden of Sir Charles Wager in England provided the material illustrated. Catesby called this tree the "Laurel Tree of Carolina":

> "What much adds to the Value of this Tree is, that it is so far naturalized and become a Denizon to our Country and Climate, as to adorn first the Garden of that worthy and curious Baronet, *Sir John Colliton*, of *Exmouth* in *Devonshire*, where, for these three Years past, it has produced Plenty of Blossoms…" "Their Native Place is *Florida* and South *Carolina*, to the north of which I have never seen any, nor heard that they grow"(II-61).

Magnolias are considered a basal flowering tree. The paleontological record goes back 80 to 100 million years, well into the age of dinosaurs. The flower, with the parts arranged in a spiral, is considered primitive. Today the southern magnolia occurs widely throughout the south. Its flowers are white, showy and large. It grows best along streams and near swamps in moist, fertile soils. *M. grandiflora* grows rapidly from seed and is an especially handsome landscape tree. The fruit (Plate 1, pg. 25) is reddish-brown with bright red kidney-shaped seeds. Once on the ground the seeds are eaten by squirrels, opossum, quail, and wild turkey.

The sweetbay, *Magnolia virginiana*, was one of the first trees introduced to Europe from North America, probably through the Reverend John Banister (c. 1650-1692), well known to English

PLATE 1
SOUTHERN MAGNOLIA
(*Magnolia grandiflora*)

M. grandiflora is a pyramidal, evergreen tree that often reaches 80-90 ft. Common names include bull bay, great flowering magnolia and southern magnolia. The oval, 6-10 in. shiny, leathery, green leaves are often rust-colored below. The 8-10 in. creamy white, fragrant flowers are composed of two whorls of visually identical thick, concave tepals. This arrangement is indicative of their ancient lineage: members of the magnolia family are considered "primitive" by botanists. Flowers are produced in the spring in South Carolina, but in cooler places the southern magnolia blooms sporadically throughout the summer and fall. The fruits are aggregated into an egg-shaped cluster, 3.5" in diameter, each individual fruit containing a single, red-coated seed that, when released, dangles on a silken thread.

This magnolia grows in coastal areas from Long Island, New York south to Texas and has escaped from cultivation elsewhere. There are many cultivars. The largest reported tree in the United States today grows in Smith County, Mississippi. It is approximately 122 ft. tall.

Catesby called this tree "*The Laurel Tree of* Carolina":

> "The Petals are usually ten, and sometimes eleven and twelve in Number; they are thick and succulent, in the midst of which is placed the Ovarium, closely surrounded with Apices, which, before the Petals fall off, swells to the Size of a Pigeon's Egg, and when fully grown, is formed into an oval Cone, in Size of a Goose's Egg" (ll-61).

The watercolor was painted from a tree growing on the bank of the South River, near Annapolis, Maryland in the fall.

collectors and gardeners, who came to the Virginia Colony in 1678. The Oxford-educated Banister was an energetic collector and adapted quickly to his new environment. From his base in Bristol parish at the mouth of the Appomattox River, he began immediately to prepare a catalogue that became the first printed survey of North American plants. He sent back dozens of specimens and seeds, but was especially drawn to the food plants - watermelons, maize (now often used as decorative Indian corn), and beans - grown by the Indians. All of these were introduced into European gardens. He created a successful estate of nearly 2000 acres manned, of course, by slaves. Banister died at age 42, the victim of an accidental shooting while exploring for plants.

Banister was interested in animals as well as plants. In 1679 in a letter to one of his Oxford professors he remarked on the abundance of passenger pigeons and of wild turkeys that weighed 30 pounds! He observed that the flesh of the turkey buzzard (turkey vulture, *Cathartes aura*) was used locally to treat the "French Disease" (syphilis) and commented on keeping a "Humming Bird" several days in captivity. Hummingbirds are endemic to the New World and were then almost unknown in England at the time. But it was the plant materials received by Banister's sponsors that were most exciting. English gardeners have been especially successful with magnolias as specimen trees.

Sweetbay is widely distributed along the coast from east of the Mississippi north as far as New Jersey and Long Island, New York. Outside of Florida it grows to about 30 ft. tall, is never far from moist soil, and prefers lower elevations. In undisturbed areas it is associated with red maple, black gum, sweet gum, water oak, and laurel oak. One interesting feature is that sweetbay exhibits two growth forms. One, roughly distributed from North Carolina south, typically has a single trunk and is evergreen. The other, from North Carolina north, is deciduous and often produces multiple trunks. However the leaves, flowers, and anatomical details are the same. Some intergradations of the two forms exist where they overlap, but identification in the wild is extremely difficult, and considered by some to be impossible. The dilemma is not resolved by herbarium specimens. While not a problem to the trees, which do very well despite their different growth patterns, it is not clear if these forms are simply geographical variants or nascent species. The dilemma is based in large part on questions regarding the genetic organization of populations within species. It has been a problem of interest to geneticists, taxonomists and evolutionary biologists. The debate revolves in part around what role intraspecific taxa play in the biology of the species and the mechanism of Darwinian selection. Small populations such as those of the sweetbay may represent adaptation to very local conditions and because the traits are heritable, encapsulate the evolutionary future potential of the species. Alternatively, they may be simply an expression of genomic variation capable of reproducing and essentially neutral in an evolutionary sense.

It is common knowledge that local conditions of soil, moisture, temperature, and light, influence plant form. The variation in form of sweetbay is most likely environmentally related. Despite this amazing array of form and life style, all plants face the same problem: capturing light and fixing Carbon Dioxide (CO_2). To be successful they must do so better than their neighbors. One measure of success is that they leave more progeny. But reproductive success involves more than competition based on photosynthesis. Other features such as the patterns, timing, and mechanisms of growth, pollination, seed maturation, fruit dispersal, seed survival and viability all contribute to the outcome of the next generation. All these aspects of heritable variation in individuals are acted on by selection; the fittest survive.

Ivory-billed Woodpecker
(extinct)

The last scientific study of the ivory-billed woodpecker (*Campephilus principalis*) (Plate 2, pg. 29), designated "*Picus maximus rostro albo*" by Catesby, was carried out in the 1930s in a cypress swamp called the Singer Tract in Madison Parish, Louisiana. James Tanner (1914-1991) documented the natural history of the species for his doctorate at Cornell University. Tanner was one member of a group from Cornell that included Arthur Allen, Peter Paul Kellogg and George M. Sutton. All were prominent ornithologists and on the forefront of contemporary conservation studies. In Catesby's time the ivory-billed woodpecker was probably relatively common, inhabiting moist, old-growth hardwood and cypress swamps, typical of bottomlands in the Gulf States and coastal Carolinas. These forests were susceptible to insect infestations. The damaged trees provided food and nest sites for these very large woodpeckers. Catesby's illustration has the bird poised on a willow oak; a plant he noted was limited to "low moist land."

By the early 1940s the number of birds, while never accurately established, had begun to decline. Tanner resurveyed the Singer Tract and explored river bottoms in South Carolina including the Santee Swamp and Peedee Rivers frequented by Catesby. Appropriate areas in Georgia, Florida, and Mississippi were also canvassed. These forests, which had supported the highly mobile search style of these large birds, were in steep decline. The disappearance of vast mature timber stands due to logging advanced so rapidly that Tanner last saw an ivory-billed in 1941. Two years later, in late 1943, a single female was found on the Singer Tract by Richard Pough, then working for the National Audubon Society. Don Eckelberry, the noted illustrator, painted the same individual in April 1944.

In his description Catesby noted that the bill, which is ivory white, was used as decoration. "The Bills of these Birds are much valued by the *Canada Indians*, who make Coronets of 'em for their Princes and great warriers, by fixing them round a Wreath, with their points outward" (1-16). As *C. principalis* did not occur much further north than the Cape Fear River in North Carolina, Catesby's note about use of the bills or skins as far away as Canada suggests they were trade items among Native Americans. Catesby, of course, was aware of the potential resources represented by the river-bottom forests. He could not, however, conceive of the devastation and ecological changes that commercial logging would bring. The fate of the ivory-billed woodpecker was intimately connected with the fate of the old-growth, bottomland forests of the Southeast. Catesby's illustration ultimately became the type specimen for the species and was cited by Linnaeus.

The ivory-billed woodpecker spent little time on the ground. It fed and bred in the middle to upper levels of dead or damaged trees, using its powerful bill to strip bark or tunnel after the larvae of large wood-boring beetles. The larvae were supplemented by fruits, nuts and berries. Overall, coupled with its typical alarm call, it was an imposing presence. Observers are reputed to have responded with a 'Lord God!' when it appeared. The attribution exists today.

The story of the ivory-billed woodpecker did not end in 1944. Every few years or so there

Ivory-billed Woodpecker
(*Campephilus principalis*)

The ivory-billed woodpecker averages 20 in. long with a 30 in. wingspan. Common names include great god bird and holy grail bird, both supposedly exclamations uttered when the bird was first seen. The head, back and wings are black. The males have a large red crest and white stripe from the beak to back of the shoulders and white wing patches. The thick beak, made of bone, is ivory white, thus the common name. The female is similar but smaller and lacks the red crest. Feet are ochre colored.

Historically, ivory-bills were distributed from the Carolinas to Florida, west to Texas and north to Illinois with reports from Cuba. Preferred habitats included mature forests with trees disturbed by insect infestations or storms. Major foods included insects and their larvae, harvested with their long strong bill by ripping bark from infested trees. Their diet also included fruits and other vegetation. Catesby wrote of this bird:

> "These Birds subsist chiefly on Ants, Wood-worms, and other Insects, which they hew out of rotten trees; nature having so formed their Bills, that in an hour or two's time they will raise a bushel of chips; for which the *Spaniards* call 'em *Carpenteros*" (I-16).

This watercolor was drawn from a specimen made available by Joel Cracraft, Lamont Curator of Ornithology at the American Museum of Natural History. Below the drawer where this specimen is lying beside twelve or so other males, is a drawer with twelve females, all shot by a collector on the same day in 1911.

are claims of someone hearing the distinctive hammering of its bill or of an isolated sighting. Unpublished photographs are rumored to be circulating, or an isolated feather turns up to attract the interest of birdwatchers and the media. A report by a naturalist in the 1950s stimulated The National Audubon Society and an interested landowner to open an ivory-billed woodpecker refuge, which closed two years later. The most recent episode (2004) in the saga occurred in Louisiana and areas of Arkansas and involved ornithologists from Louisiana State University and the Cornell Laboratory of Ornithology. On a report from a reliable source of hearing a bird in this area, a thorough search was initiated. No bird was seen or heard, but voice-activated listening devices were distributed in the area. Even after several years, no birds were detected and the search has been abandoned.

The tale of the ivory-billed is the story of the forests. The forest wetland or bottomland was the habitat most familiar to Catesby. The ecosystem included a wider variety of tree species and a greater range of tree ages than exist today. The trees were mature and mast crop production was high. There were not only big and old trees, but decaying and fallen trees that increased the variety in the habitat. Such conditions of tree variety are precisely those that favored both diversity and large populations. Little, if any, of this primal forest exists today. Pre-colonial bottomlands were dominated by oaks (*Quercus*), gum (*Nyssa*), and cypress (*Taxodium*). Catesby observed that, "The Soyl of *Carolina* is various, but that which is generally cultivated consists principally of three Kinds ..." (I-iii) and recognized that the composition and moisture of the local microclimate causes a patchy distribution even within the dominant forest type. So cypress-tupelo, overcup oak (*Q. lyrata*), and water hickory (*Carya aquatica*), inhabit the wettest areas. Sweetgum (*Liquidambar styraciflua*) and water oak (*Q. nigra*) occur in areas of intermediate soil moisture and perhaps some standing water. The familiar hickory, oak, and magnolia thrive in the drier, and still flat, tidewater interior.

This distribution changed over time as disturbances occurred and the composition of the forest changed with the turnover in species. Tree species vary in rates of growth, tolerance to moisture and light conditions, or dispersal abilities. Which plants invade and settle depends on what is already in the local and adjacent areas or is introduced naturally. Many trees and shrubs, in most ecosystems, produce seeds that are dispersed by birds. Hence, the long term composition of the plant community may depend on the diversity of the resident bird species. The cycles run on time courses measured in dozens of years or even perhaps decades.

Reverend Bachman's Warbler
(extinct)

Catesby noted over 100 species of birds, 71 of which were new to science, thus needing names. The naming of species, especially the common names, can be both tortuous and fascinating. A species is assigned a name by the person who first describes it scientifically. This may not be the same individual who found or collected it. There are rules and regulations for scientific names established by international groups, but not common names. This insures some consistency to the process, and aids immeasurably in subsequent retrieval of the information. It usually does not impinge on the creativity of the authors for common names. The history of taxonomic nomenclature, accordingly, is filled with puns, patriotism, sentiment, and, to some, chicanery. One coincidence, probably not unprecedented, is that the last bird Audubon described, Baird's sparrow (*Ammodramus bairdii*) was the first to be named after Spencer Fullerton Baird (1823-1887). Baird, an outstanding nineteenth Century ornithologist, had a long career at the Smithsonian Institution. He introduced himself at the age of 17 to Audubon, who subsequently collected the sparrow specimen.

The bird Audubon named Bachman's warbler, *Vermivora bachmanii* (Plate 3, pg. 35), was collected near the Edisto River just south of Charleston. John Bachman (1790-1874), a Lutheran minister who headed the congregation at Charleston's St John's Church, secured the specimen. Bachman had a longtime interest in birds. In the summer of 1804, as a schoolboy in Philadelphia, he had collected three specimens of an unusual jay. Alexander Wilson, a local ornithologist, identified them as Canada jays (now gray jay, *Perisoreus canadensis*), the first collected in the United States. Bachman accumulated a moderate-sized collection from the area around Charleston, much the same ground covered by Catesby a century earlier. Catesby did not illustrate this warbler but did paint the yellow rump warbler (*Setophaga coronate*), hooded warbler (*Setophaga citrina*), pine warbler (*Setophaga pinus*) and yellow-throat warbler (*Setophaga dominica*). Bachman kept meticulous field notes on birds regarding diet, movements such as migration, and the properties of nests and eggs. Documentation of this type, unfortunately, was not a priority for Catesby. Catesby collected with a different purpose. He had been engaged by his sponsors to survey, collect, and return materials to England. Bachman and others of the period were sophisticated hobbyists interested in local natural history. They usually had day jobs. Audubon was not typical in that he attempted to support himself through his collecting and publications.

Bachman arrived in Charleston in 1815, married Harriet Martin and started a family. By 1831 when Bachman first hosted Audubon, his family consisted of six daughters and eight sons. Maria Martin (1796-1863), Bachman's sister-in-law, also lived with the family. In addition to helping with the family, she was an accomplished biological illustrator. Not only did she participate in collecting and preparing specimens for Bachman's "cabinet," she eventually became his wife after the death of her sister. In the interim, she illustrated plants for many of Audubon's bird plates, and the birds in at least one.

Bachman had met Audubon earlier when he had spent time in Philadelphia. Two of Bachman's daughters eventually married Audubon's two sons. In addition to this link through their children, Bachman and Audubon cooperated scientifically. Audubon had named another species 'Bachman's Pinewood-Finch'. He did not realize, however, that it had been described previously from a specimen taken in Georgia. Nevertheless, this species is now known as Bachman's sparrow (*Aimophila aestivalis*). Audubon also attached Bachman's name to the black oystercatcher (*Haematopus bachmanii*) a species from the Pacific Coast of North America.

Bachman and Audubon almost certainly went into the field together, as Audubon visited Charleston often. In addition to his observations on birds that were incorporated into Audubon's *The Birds of America* (1827-1838), Bachman provided the text for Audubon's landmark *The Viviparous Quadrupeds* (1845-1854). The family relationships became deeply intertwined. Both of Bachman's sons-in-law were involved in editing and illustrating various publications. Bachman and Audubon also participated in the growing circle of naturalists who corresponded, shared experiences and specimens, and entertained visiting Europeans.

In addition to his knowledge of the birds and mammals of North America, Bachman thought extensively about the nature of species - not necessarily their origin, but certainly their boundaries, variability, and priority in their description. Later in life, Bachman became embroiled in the issue of the unity of *Homo sapiens*. Luminaries such as Professor Louis Agassiz (1807-1873) of Harvard held that each of the human races (Africans, American Indians, Asians, etc.) was a separate species. Bachman argued for the unity of mankind based on religious beliefs and on biology. Because, as was believed at the time, all humans were descended from Adam and Eve, it was hard to argue otherwise. Bachman's position was strengthened based on his extensive knowledge of intraspecific variability. He was also aware that animal species rarely interbred and that hybrids were frequently sterile. A common misconception at the time was that most new species arose by hybridization. Clearly, human interracial crosses were not sterile and hybrids were not new species. Beyond the biological implications, this argument had important political overtones, especially in the slave-holding southern states. Bachman remained confident that humans were all one species. Politically, this was a difficult position to hold in the Antebellum South, where the issue of slavery and the scientific question of the unity of humans were being tested. He welcomed Darwin's *On the Origin of Species* (1859). A year later Lincoln was elected president and South Carolina seceded from the Union. After extensive debate and considerable soul searching, Bachman eventually supported the Confederacy. John Bachman outlived Audubon and several of his own children. He died in 1874.

Bachman's Warbler is now probably extinct. Like other warblers it was a small, colorful, active, insect-eater. The plumage was sexually dimorphic and was replaced seasonally. This situation is not uncommon in birds, and results in four distinct plumage classes, excluding the juvenile plumage in first-year birds. That is, there are differences between the sexes and there is, at least, a basic and alternate (nee breeding) plumage for each sex. Bachman's warbler migrated from the Carolinas to Cuba for the winter, and was associated with swamps from its discovery. The disappearance of Bachman's warbler has to do with changes in its niche: the place where it makes its living.

Settlers in America have altered the environment in many ways, perhaps most dramatically through clearing land. Species dependent on habitat with high commercial value, such as old growth forest, are extirpated along with the harvest. The ivory-billed woodpecker was dependent on the mature floodplain forest of the southeast. Harvesting and disturbance such as fire and periodic flooding affected the forest understory and the alluvial floodplain. This generated thick, extensive stands of giant bamboo (*Arundinaria*) or cane, the canebrakes of the southern swamps.

Cane, one of two native bamboos, grows in areas subjects to seasonal flooding and fire. Ironically, as the forest was removed the canebrakes expanded, as did Bachman's warbler. The bird nested, perhaps exclusively, in the bamboo. Late nineteenth-century naturalists noted the increase in numbers even though little was known about nesting behavior, as the cane was so impenetrable and few nests were ever found. William Bartram described the canebrake as "…ten or twelve ft. in height, and as thick as an ordinary walking-staff; they grow so close together, there is no penetrating them without previously cutting a road." Further, this "vast plain" extended as far as the eye could see. Formidable indeed! Subsequently, when the rich alluvial floodplain was drained for agriculture, and fire control measures were introduced in the remaining woodlands, the canebrakes disappeared. The final decline of the warbler was probably abetted by changes on the wintering grounds. Forestlands in Cuba were cleared for farming in the 1920s and the warblers extinction may have been hastened by a series of hurricanes in the 1930s. The combination of human activities and natural phenomena was overwhelming.

Bachman also discovered Swainson's warbler (*Limnothlypis swainsonii*), which is now considered a federally threatened species and is endangered in at least one state. *Limnothlypis* is also a creature of dense bottomland forest, swamps with dense undergrowth, and canebrakes. With habitat requirements similar to those of Bachman's warbler, its increasing scarcity is a direct consequence of continued habitat destruction. While William Swainson (1789-1855) had nothing to do with its discovery, Audubon admired his work. Audubon also described Swainson's hawk (*Buteo swainsonii*). Swainson, an Englishman, was an accomplished naturalist, prolific writer and artist. He served in the Napoleonic Wars, traveled and collected widely, and sometimes interacted tumultuously with Audubon. His personal life was a series of misadventures and disappointments, with a number of unfinished projects on birds. He died in New Zealand.

Bachman's Warbler
(*Vermivora bachmanii*)

Bachman's warbler, also called the swamp warbler, was approximately 4 in. long with a wingspan of 4 1/2 in. The slender beak was curved slightly downward. Males were olive-green above, with a yellow forehead, eye-ring, chin, and underparts, and a black throat and crown with yellow shoulder patches. The female had olive green upperparts with yellow forehead and grayish crown. The pale white eye-ring was more noticeable in females than in the males.

This extinct bird bred in the southeastern U.S. and wintered in western Cuba. It nested near the ground in flood-plain forests amongst the thick groves of cane. Its diet included insects and small invertebrates.

I was surprised at the diminutive size of the pair, and the subtle coloring.

The specimens for the watercolor were from the Yale Peabody collection through the courtesy of Professor Rick Prum, William Robertson Coe Professor of Ornithology, at Yale University, New Haven, Connecticut.

Arundinaria gigantea, known as American bamboo, giant cane and river cane, is a native woody perennial grass. It grows from thick rhizomes forming dense colonies. The stems are 3-25 ft. tall, at first unbranched, later branching and forming fanlike clusters with upper leaves longer than those below. This grass flowers infrequently. A smaller form grows 6-8 ft. tall and prefers wet soil.

PASSENGER PIGEON
(extinct)

In 1778, as he traveled from Savannah to Charlestown, William Bartram was an overnight guest in the home of Jonathan Bryan. The gardens on the property were delightful and Bartram was warmly welcomed. He remarked that:

> "At night, soon after our arrival, several of his servants came home with horse loads of wild pigeons (Columba migratoria), which it seems they had collected in a short space of time at a neighbouring Bay swamp: they take them by torch light: the birds have particular roosting places, where they associate in incredible multitudes at evening, on low trees and bushes, in hommocks or higher knolls in the interior parts of vast swamps. Many people go out together on this kind of sport, when dark: some take with them little fascines of fat Pine splinters for torches; others sacks or bags; and others furnish themselves with poles or staves: thus accoutred and prepared, they approach the roosts; the sudden blaze of light confounds, blinds and affrights the birds, whereby multitudes drop off the limbs to the ground, and others are beaten off with the staves, being by the sudden consternation, entirely helpless, and easily taken and put into the sacks" (Bartram, 1791).

Bartram noted that it was the mast crops of the abundant oak trees in the area "…which induce these birds to migrate in the autumn to those Southern regions: where they spend their days agreeably and feast sumptuously . . .", He said nothing regarding the gastronomic aspects, but clearly Passenger Pigeons were a staple of the diet throughout the colonies. The ease with which the birds were collected and their abundance would provide fodder for the tables of many families for decades to come. Given the remarkable supply, the birds were sold at market for often less than half-a-dollar a *dozen*! Unfortunately, these were precisely the same conditions that would lead to the extinction of the species.

Catesby, with his usual acumen, illustrated the passenger pigeon (Plate 4, pg. 39) in a red (turkey or scrub) oak. He noted: "Of these there come in Winter to *Virginia* and *Carolina*, from the North, incredible Numbers…". They are so numerous in places that "they often break down the limbs of Oaks with their weight, and leave their Dung some inches thick under the Trees they roost on." Further, they "…so effectually clear the Woods of Acorns and other Mast, that the Hogs that come after them, to the detriment of the Planters, fare very poorly" (I-23). This is a remarkable biological force attributable mostly to their vast numbers. Linnaeus used Catesby's information for his original scientific description of the passenger pigeon.

As early as 1709, John Lawson described the species as so numerous it would take a flock a quarter of an hour to pass, and that flocks had continued to fly by for an entire morning. Catesby reported that "…I have seen them fly in such continued trains three days successively, that there was not the least Interval in loosing sight of them, but that some where or other in the Air they were to be seen continuing their flight South." It is difficult to appreciate the numbers of these birds and the biomass they represented. They were abundant to the point where Catesby mentioned that "…the People of *New-York* and *Philadelphia* shoot many of them as they fly, from their Balconies and Tops of Houses; and in *New England* there are such Numbers, that with long Poles they knock them down from their Roosts in the Night in great numbers" (1-23).

Despite the heavy toll, the pigeon's numbers held up for some time. A century after Catesby's visit, Alexander Wilson, one of America's most important ornithologists, described a nesting site in Kentucky that was: "…several miles in breadth and over forty miles wide. Single trees may have contained more than 100 nests, each with a single egg" (Cokinos 2000).

He described one flock which he estimated at 2.2 *billion* birds. Passenger pigeons were perhaps the most numerous land vertebrate prior to the arrival of Europeans and may have reached a biomass that approached the American bison. It had been reported that even in 1874 in Michigan alone birds were taken at the rate of perhaps 700,000 a month. Just as descriptions of the numbers and information on their breeding became known, so did new ways to cook them, use carcasses for fertilizer, and sell their parts and products as folk medicine. As farming spread westward, the birds presented a different type of problem: crop damage. Farmers and market hunters continued to shoot and net birds, often in outrageous numbers. Pigeon shoots became a popular recreational activity. The techniques used on the Bryan Plantation 150 years earlier were supplemented by shooting, netting, and trapping.

The party could not last for ever. Audubon, in the early nineteenth century, realized that the number of passenger pigeons was in decline but probably did not fully comprehend the long term consequences as the birds remained numerous in his time. By the 1890s it was unmistakable that the numbers were in a steep decline. Hunting had taken its toll, and in the west the forests with their abundant mast crops had been cleared by settlers for farming. It is also possible that canker (trichomoniasis, *Trichomonas gallinae*) acquired from the introduced rock pigeon (*Columba livia*) contributed to the passenger pigeon's decline.

Laws to protect the birds came first in states in the mid-west, but it was too little, too late. The last wild bird was shot in 1900, and the last of the species, Martha, a captive bird in the Cincinnati Zoo, died in 1914. She is reported to have been named for President George Washington's wife. Martha's demise marked the extinction of both the species and the genus. Martha had become something of an icon. Her skin was prepared at the Smithsonian and it has appeared in various displays in support of conservation. The Charleston Museum has a single display dedicated to the passenger pigeon, Ivory-billed woodpecker, and Carolina parakeet (*Conuropsis carolinensis*), all set in a diorama of a cypress swamp.

The extinction of the passenger pigeon, considered a key species in the eastern forest ecosystem, had additional consequences. It is possible that due simply to their large numbers the passenger pigeon controlled the mice numbers by out-competing them for food, in this case the acorns from the wide-spread oak trees. Both mice and deer provide a reservoir for *Borrelia burgdorferi*, the bacterium that causes Lyme disease in humans. As larvae, ticks in the genus *Ixodes* feed on mice and other small rodents. The adult ticks usually feed on deer, hence the reference to deer ticks. As the numbers of passenger pigeons plummeted, more resources became available for the mice, and their numbers increased. Deer numbers eventually increased as well when their natural predators were eliminated and their habitat was protected. With humans moving

Passenger Pigeon
(*Ectopistes migratorius*)

The male passenger pigeon measured approximately 16 in. long with a wing span of 16-18 in. Females were an inch smaller. The extended 7-8 in. tail, along with the long tapered wings, increased the capacity for great speed, and gave this dove the moniker 'Blue Meteor.' The head and neck of the males were an iridescent bluish gray with black streaks on the scapulars and wing coverts. Iridescent feathers at the sides of the throat and back of the neck changed color with the light. The breast faded from pink to white on the abdomen. The eyes were red and the eye-ring purplish. Legs and feet were also red, the bill black and pointed. The female colors were softer and paler.

Passenger pigeons once ranged from Ontario to Nova Scotia, south through Texas, east to Florida. They nested from the Great Lakes to upper New York. They were found in mixed hardwood forests. The diet was beech nuts, acorns and chestnuts or insects and worms when available.

Martha, the last surviving passenger pigeon, had lived for 29 years in the Cincinnati Zoo. Upon her demise, she was prepared by taxidermists and displayed at the Smithsonian Institution, Washington DC. Twice she has been loaned to other institutions, flying first class with a flight attendant as chaperone. I am sure the 'Blue Meteor' would approve.

I painted this species from specimens in the American Museum of Natural History, New York.

into previously forested areas, in time, the association between people and deer became more intimate. Lyme disease was first recognized in 1975 in Lyme, Connecticut, but was probably present earlier in the twentieth century.

Since 1800 the number of known bird species has increased to about 10,000. In this same period about150 species are known to have become extinct. The rate of finding new species has varied over time, as have extinction rates. Newly discovered species have higher extinction rates than those known for centuries as they are more likely to occur in restricted, isolated habitats and in small numbers. The earliest species described tend to be widespread and abundant, and hence less prone to extinction. On oceanic islands extinction rates tend to be highest in the years following human settlement.

Certainly, species of plants and animals have arisen and gone extinct in the past without human knowledge or human interference. There are examples of others that have disappeared as a consequence of habitat destruction, only subsequently to be rediscovered. Predictably, as habitat restriction and destruction proceed, species will continue to disappear. While small populations may linger in a reduced patch of habitat before expiring, perhaps an additional 1,250 species will be lost in this century. With the acceleration in the rate of world-wide habitat loss, the additional loss from introduced predators and disease, and the unanticipated detrimental effect of a new human activity, these numbers are only estimates. In the near future, the most potent factor in species extinction is likely to be climate change. One estimate holds that some 15%-37% of species in existence today are at risk in the next 50 years from human-induced climate change.

RICE

The origin and domestication of *Oryza sativa* are unknown. Archeological evidence point to a time, perhaps 8,000 years ago, when rice was first cultivated. Rice has fed more people over time than any other crop. A second species, *Oryza glaberrima,* was domesticated in interior Africa 3,000-4,500 years ago and spread to the Atlantic coast of Africa before the arrival of the first Europeans. Because of its higher yields and ability to withstand milling, Asian rice was favored economically and diffused more widely. Rice (Plate 5, pg. 43) has been used in brewing beer, paper making, and served as a side dish for fish on meatless Fridays. Portuguese explorers in 1446 discovered African rice on the west coast between Senegal and Liberia. Rice, along with millet and sorghum were used to provision caravels as they explored the African coast and routes to spice-rich Asia. In the sixteenth century the Cape Verde Islands became a crucial trading port for commerce and an important stop for the Portuguese on the route to West Africa and India. African slaves on the islands raised crops to feed the merchant vessel crews and by 1514 rice was aboard ships bound for South America. Rice was introduced to South Carolina prior to 1700, probably by slaves brought by English settlers.

Catesby commented that Sir Nathaniel Johnson was only marginally successful in growing rice in the Carolina Colony. There is an old story, not fully documented, that in 1685 a distressed merchant ship put in at Charleston for repairs carrying rice from Madagascar. Slave-ship masters regularly provisioned their ships on the African coast and favored *O. sativa* as it was less susceptible to moisture damage in passage and the unhusked rice could serve as seed for planting. The cargo in this case originated from the windward coast of Africa, most likely Sierra Leone, and included slaves who brought with them the rice grown at home. This was likely only one of several introductions of rice to the Americas. Its serendipitous arrival as the better grain intersected with the technological knowledge of the slaves and nearly ideal local conditions for its cultivation. In time, growing the crop in the Carolinas would produce enormous wealth, but ultimately would also devastate a portion of the environment. And of course the issue of slavery in America eventually led to a horrifying war.

Rice requires unusual conditions for growth. Water is critical - rice grows well in regions of monsoon rains or swamps irrigated with fresh water. On the African floodplains a farming system had developed in estuaries that used marine tides in cleared mangrove swamps. The tides kept the soil moist prior to cultivation and then rainfall served to desalinate and irrigate the crop. Growing rice in swamps represented a major achievement of the indigenous rice culture. It was the combination of the skills of the slaves and the characteristics of the rice that built an entire industry in South Carolina.

Moderate rice crops had been grown in the Carolina inland swamps since the late 1600s, but it wasn't until 1750 that the planters began to dike the marsh flatlands. At the time, the War of Jenkins' Ear (1739-1748) had cut off the market for indigo, a plant dye that was the colony's most lucrative export. The introduction of dikes modified the use of the fields and prepared the way for an heir to that crown. The flooding tidal waters provided the cropland suitable for the swamp rice with nutrients and also purged insects that could potentially damage the plants. The water movement had to be gentle enough not to wash out the seeds, or disturb young plants. The slaves from Sierra Leone had long ago perfected the planting of seeds by placing a grain of rice in a small ball of mud and leaving it to dry. These 'plugs' were inserted into the fields by hand. The low country of South Carolina is ideally suited for this style of rice cultivation. Rainfall is high and numerous fresh-water rivers, streams and creeks cross coastal marshes to form broad river deltas. The tidal cycles of the Western Atlantic provided the power to drain and fill the marshes. Work began with clearing trees and building dikes, both of which took place in knee-deep water, often shared with snakes and alligators. Once the swamps were cleared (and these were magnificent old-growth bald cypress swamps), dikes were constructed around what were to become the rice fields. Double gates controlled the flow of water in and out with the tidal cycle. Because seawater is denser than fresh, the filling tide lifts the water in the channels into the fields; with the ebb, fresh water filters through the soil. With careful management, the sea-water never reached the crucial top layers of soil: salt water would destroy the rice. The dikes and gates, all hand built, were under the watch of the 'dike masters,' critical members of the plantation community. Tidal cultivation required heavy inputs of labor and capital, but one slave could produce five to six times the yield of the previous methods of growing rice.

The planting and harvesting of rice, the region's dominant crop in this period, was extremely labor intensive. Governor Johnson told Catesby that he "…had procured from Spain a machine which facilitates the work with more expedition…", but there is no record of its success (Frick and Stearns, 1961). From the earliest times, slaves performed the drudgery of planting and

Rice
(*Oryza sativa*)

O. sativa is a semi-aquatic annual grass with several jointed culms or stems from 2 – 5 ft. tall. The internodes are hollow. Each node bears a leaf and bud, which may grow into a shoot or tiller. The inflorescence, called a panicle, is born at the tip of the culm. The harvested kernel is enveloped in a hull or husk known as rice bran.

Catesby drew this rice with the rice bird or Bobolink. He wrote:

> "*Anno* 1724, an Inhabitant near *Ashley* river had forty acres of Rice so devoured by them, that he was in doubt, whether what they had left, was worth the expence of gathering in" (I-14).

I observed and photographed this rice growing at several South Carolina plantations and finished the watercolor by using wild rice as a model. The material was gathered for me by Carol and Gary Weed from the Connecticut River in Lyme, Connecticut.

harvesting the crop. W. E. B. DuBois estimated that 14 million slaves were imported during the run of the rice industry. Slaves planted, maintained, and harvested the crops through the hot, humid growing season. Unfortunately, rice fields also provided an ideal breeding place for mosquitoes bearing malaria and yellow fever. Childhood diseases such as smallpox, measles, whooping cough and dysentery were common. These were the major causes of death in both the black and white populations of the region. Parts of South Carolina were truly the swamps of summer.

During the early years of the rice industry, slaves formed the majority of the population in South Carolina, and they were considered an important asset of the plantation system. The dilemma the owners faced was that the drivers and dike masters were often skilled managers and well trained to do their job. It was natural to favor such individuals, but often illegal. This violated the law because these favored slaves may have been taught to read and write (illegal), to occupy responsible positions (illegal), and replaced the use of a white overseer (also illegal). In any case, growing rice in coastal swamps was a highly profitable venture.

In 1720-1726, roughly the period of Catesby's Carolina visit, there were 6,250 white settlers and 10,500 enslaved blacks in the colony. The Charleston area annually imported 600 slaves and exported 71,000 hundredweight (1 hundredweight= 112 lbs.) of rice. From 1731-1738, an average of 2,000 slaves were imported annually and 143,000 hundredweight of rice exported. By 1740, about 40,000 slaves were estimated to be living in the area, and annual rice exports exceeded 143,000 hundredweight, about three-quarters of the colony's production. Plantations of more than 10,000 acres were established along the Santee and Waccamaw Rivers. Even larger holdings developed along the Cooper and Ashley rivers. By 1848 the Charleston area produced 15 million pounds of rice annually. Of 559 plantations that produced over 20,000 pounds per year, 446 were in the South Carolina tidal marshes. By 1860 a dozen planters produced a million pounds each. Rice was marketed in Charleston and shipped worldwide.

Selective breeding improved the rice until 'Carolina Gold' became the standard. It could be grown essentially without artificial fertilizer or insecticides because of the irrigation system that supported it. But the costs in labor were enormous and the economy rested entirely on slaves. Great fortunes were created in the century from the mid-1700s to the Civil War. The run ultimately ended with President Lincoln's emancipation of the slaves in 1863. The final blow came from nature when a series of hurricanes just before and after the Civil war forced sea water into the rice beds. There was no cheap labor to rebuild the plantation system and little capital to finance recovery.

The story of the rice-based economy was remarkable. Fortunes made in rice provided a foundation for real estate that often included the plantation house and satellite structures, an elegant home in Charleston, and seasonal retreats in the mountains or on the shore: even as distant as Newport, Rhode Island. Owners and their families moved seasonally from one site to another and even traveled to Europe. Besides travel, many maintained an extraordinarily active social life. The landed gentry participated in colonial governance and, after the revolution, in both state and national politics. Grown children were privileged and it was thought especially honorable for young men to serve in the military, as it was to attend college, enter a profession, or the ministry. The social system and economy of this period were constructed on changes in land use and the labor of slaves. When threatened with radical changes to the slave-based system, in 1861 South Carolina was the first state to secede from the Union.

In hindsight, it is clear that the planting and cultivation of rice carried tremendous costs. First was the human cost of slavery. Second, was the long term impact on the ecology of tidal South Carolina. Forests were destroyed, drainage patterns modified, and land use patterns changed. At least two bird species were started on the road to extinction. Eventually, both plantation owners and their slaves left the low country in favor of other states with richer prospects in cotton or other crops.

Today, rice is a staple in the diets of over half of the world's population. Other grains, wheat for example, make significant contributions to human nutrition, but none feeds as many as rice. Over evolutionary time grasses have been reliable food sources for many animals, from sparrows and small rodents (e.g. mice and voles) through all manner of open-land browsers, such as bison and elephants.

The cultivation of grains has also had profound environmental effects. One example is the degradation of soil quality and erosion of the grasslands of the North America Midwest. The native grazers were removed and the once diverse grasslands replaced by a crop monoculture overly dependent on irrigated water. In time this combination led to the dust bowl catastrophe of the 1920s and 1930s.

Seeing The Forest and The Trees

The American forest, as it was before Europeans arrived, now exists only in the imagination. Present-day environmentalist David Wilcove imagines travel in pre-industrial times, around 1498, in the Mid-Atlantic States as:

> "…a major expedition, for we must proceed on foot, taking advantage of whatever game trails or Indian pathways we find. For most of the journey we travel within a magnificent old-growth forest. Oaks, chestnuts, and beeches tower above us. Beneath them are two or three layers of smaller trees, while jack-in-the-pulpits, may apples, and other wildflowers cover large portions of the forest floor. The songs of red-eyed vireos, black-and-white warblers, scarlet tanagers, and other songbirds fill the morning air, but the birds themselves are surprisingly difficult to spot in the tall trees. In wet mud along the Patuxent River floodplain, we find the tracks of a mountain lion, as close as we shall come to spotting the elusive cat. Proceeding farther, we startle a herd of elk, glimpsing a dozen tawny rumps as they disappear within the forest. Had we begun our trip in January instead of May, we might have encountered bison, but they have since moved north, where they will remain until the snow once again brings them back to our area" (Wilcove, 1999).

Travel in Lawson's time, 200 years later, would not have been much different. Very early settlers had little apparent effect on the environment.

Franklinia

Shortly after Catesby's visit, in 1765, John and William Bartram made a collecting expedition along the Altamaha River in eastern Georgia. During their travels they discovered a stand of small trees with showy white flowers. The plant turned out to be a previously unknown member of the tea family and was named *Franklinia alatamaha*. (Plate 6, pg. 49) The name honored both the "illustrious Dr. Benjamin Franklin" from Philadelphia and the river itself. William Bartram wrote:

> "We never saw it grow in any other place, nor have I ever since seen it growing wild, in all my travels, from Pennsylvania to Point Coupé, on the banks of the Mississipi, which must be allowed a very singular and unaccountable circumstance; at this place there are two or three acres of ground where it grows plentifully" (Bartram, 1791).

Franklinia was last recorded in the wild in 1803. Catesby did not record or collect *Franklinia*, and no other wild specimens have ever been found, although the species thrives in cultivation. How and why *Franklinia* became extinct in the wild is a matter of speculation. The species exists today because of the seeds the Bartrams propagated. If other populations existed, they went unnoticed and disappeared before others found them. The number of species that disappeared as Europeans settled the area is unknown and, in essence, unknowable.

To eighteenth-century naturalists who thought about it, communities of plants and animals represented a combination of species meticulously adapted to the environment. They were as much a part of creation as the individual species; after all, God created the heavens and earth and there was no reason to suppose He didn't do so with such communities as well. Landscapes were the product of a grand design that reflected nature's overall harmony. It was presumed that individual species of plants or animals were immutable, played their assigned roles and the entire system fitted together like a huge jigsaw puzzle that ran smoothly enough to be self-perpetuating. The ecosystem, like the organisms that composed it, seemed designed for its particular role and had been since time began.

We now know this harmony of design is an illusion. That nature constitutes networks of spatial and temporal relationships based on intertwining chains of cause and effect implies a certain lawfulness, perhaps designed by a Creator. The assumption of a divine origin of nature's plan is perhaps an anthropogenic conceit based on a human socioeconomic structure. The community succeeds and profits to the extent that the participants thrive in coexistence. However, ecological communities are in fact driven by the self-interest of their constituent units at the most elemental level, in this case the species. A well-balanced ecosystem, one in an apparently steady-state, can be considered an economy, not an adaptation. The incremental organization of a community is the cumulative result of natural selection where individual components—again the species—survive because they are successful in the local conditions. Survival is the reward for selfish behavior. Plants flourish for their own good, not for the good of herbivores that consume them, the insects that pollinate them, or for the pleasure of the humans that grow them in gardens.

Ecologists for much of the twentieth century considered the landscape to be the result of successive ecological change. Once mature, a dynamic equilibrium emerged which represented the fullest development of local ecological conditions. Indeed, a sort of Darwinian selection was involved. The parts were well defined for their individual roles, and processes such as energy production could be measured. Disruptions to a community, especially those as the result of fire, drought, pestilence, or weather, would initiate a process of succession, or replacement, which would restore the equilibrium. Succession would account for the apparent long-term stability of habitats and the diversity of the flora and fauna. One implication was that the community is somehow capable of rebuilding itself and that it always follows the same pattern and will contain the same bits and pieces, the same cogs and links.

Ecological communities under this rather traditional model are viewed as an assembly of species defined by their niche. Technically they are limited-membership assemblages of species that coexist under strict niche partition of limited resources. The niche is defined as the function or role of an organism within its environment. It is the job it does. The assemblage efficiently processes the available resources, ranging from light and oxygen to chemical elements. Each species harvests what is required for reproduction, development, and growth: everything necessary to play its role. Implicit is this concept is that extinction would leave a niche unfilled; or, if left undisturbed, there would be no niche for a potentially invasive species. The community envelops a defined number of roles for plants, animals, and microorganisms. To remove one would influence the others, like pieces of a jigsaw puzzle.

Franklinia
(*Franklinia alatamaha*)

Franklinia is a small tree and a member of the Theaceae, or tea family, related to camellias. It grows 15-25 ft. with multiple stems and trunks. The leaves are oblong, serrate, 4-8 in. long, with brilliant fall color. White flowers 3 in. across with bright yellow-orange centers look like eggs sunny-side-up. The bark is gray with white striations for winter interest.

This tree was discovered in 1765 on the banks of the Altamaha River near Fort Barrington in Southeast Georgia by William and John Bartram, father and son. They collected seeds, which grew in their garden. Returning years later to the same spot, the trees were gone. No other examples have ever been found.

I grew this plant successfully in Mystic, Connecticut, for a few years but was unable to meet its requirements long-term. Fortunately for me, a property near me had two trees, now several years old, which thrive. I have seen it growing in appropriate conditions from Connecticut to South Carolina.

The processes mean also that over time the community had developed niches that elaborate on its harmonious function. More pieces in the puzzle will increase complexity, and enhance stability. The mature community derives its stability from its complexity. In this model the organization is goal-oriented. However, we now understand that selection cannot work towards a goal as it has no vision of the future.

An alternative model conceives ecological communities as non-equilibrium assemblies. Communities are open structures that undergo continuous change. The presence, absence, and relative abundance of species are governed by the random processes of dispersal, ecological drift, speciation and extinction, not the need to fill a particular niche. The approach is neutral in an evolutionary sense in that the communities do not evolve in a goal-oriented fashion. The landscape and its composition are the result of complex random events, both physical and biological. It is neutral also in the sense that all species are considered equal in their probability of immigrating into the system or going extinct once there. There are general limits given the physical features of the environment, climate and geography. This model applies equally well to the settlement of islands (see Bahamas pg. 141), restoration of landscape following violent weather events, or environmental changes over longer periods of time. It applies both to terrestrial and marine ecosystems.

Consider the bottomlands of the Carolinas. Even if left undisturbed by humans the bottomland forest will change. Flooding, which brings nutrients and soil to the forest floor, depends on the periodicity and amount of local rainfall. Soil of the lowland forest will build from the silt borne by floodwaters and be enriched by leaf fall. Flooding will influence the composition of the understory and may even affect its structure. Extended flooding during the growing season can kill trees and eliminate foliage near the ground. The time period for these changes is on the order of years and is highly contingent on the local climate.

Birds respond, too. Seasonal flooding interferes with ground-nesting species, but may create habitat for water birds. Female wood ducks (Catesby's "Summer Duck") feed preferentially on the high protein micro-invertebrates available during flood periods as they prepare to lay their eggs. More nesting birds may be attracted by insects that increase with the wet conditions. Moisture-tolerant plants thrive. Change is continual, proceeding along a continuum of time scales, and each instance can affect the diversity of the bottomlands' organisms in a multitude of ways.

Other ecological disturbances occurred after the arrival of European settlers. They, as the native peoples before them, cleared and converted the forest to other uses. In early times the wood was used for building and fuel. Human effort converted logs into structures and recovered the chemical energy in wood as heat for domestic use. Early settlers required significant amounts of lumber for building, all of which was felled and moved by hand. In addition to the tonnage required for construction, the average home in New England consumed as much as 15 cords (1,920 cubic ft.) of wood for heating and cooking annually. A village of 200 homes could deforest as much as 75 acres in one year. The cleared lands were used to grow crops and to graze livestock. Where the land was poor, it might have been fertilized by manure or simply abandoned after a few years of use. After all, more land was available simply for the clearing. The settlers' land-use interrupted the normal process of reforestation; the exception being the woodlots maintained as a future source of wood.

This pattern of land use worked well enough as an economic system until the population grew, and efficient transportation became an option. To meet changing circumstances, crops could be rotated or fields could be used to raise cash crops such a tobacco or rice in addition to what was grown for subsistence (see Rice pg. 40). Land holdings could be increased as the labor market grew. Nothing would increase productivity, however, as much as the appearance of machines or the eventual use of chemical energy sources. Both were the result of petroleum-based technology. Artificial fertilizers, rich in nitrogen, increased farm productivity and lengthened the productive life of fields. Much of agricultural chemistry depended on the commercial production of glass. Europeans used glass vessels to isolate nitrogen and then fix it as ammonia, which was then used to produce fertilizers and explosives. Machines run by coal or gasoline eventually replaced water, wind, and animal muscle.

As is well known, both the forest and the animals have changed in the centuries since European settlement. By 1872 forests in the eastern United States had been reduced to about half that present when Europeans arrived. The bison (illustrated by Catesby) and the elk retreated early from eastern woodlands. Mountain lion and wolf were hunted nearly to extinction. As consequences of forest clearing, fewer species of plants survived, less desirable plants appeared, and fewer species of animals occurred.

A reduction of the size (area) of an ecosystem usually ensures a decrease in species diversity, as the number of species is generally a function of area. For the birds of eastern North America, this reduction contributed to the extinction of the Carolina parakeet and the passenger pigeon. These two species existed in tremendous flocks in Catesby's time and were extinct by the end of the nineteenth century. Both species depended on the mast crops of trees, especially beech, oak, and hickory.

These trees dominated the upland forests of the Appalachians until change overtook the system. Hogs, introduced by the Spanish, grazed nuts on the forest floor, which reduced the reproductive success of the trees. Trees were cleared. The ecosystem was thrown out of kilter. With regeneration and re-growth, forests may recover, albeit only partly. The majestic size of trees in the native forest has not returned. Nor are the species as numerous because recruitment has been from a reduced species pool. As the land once covered by forest has recovered and its area increased, species associated with fields and pastures have declined. In parallel, forest-dwelling animals may increase but the fauna do not always return to the earlier levels of diversity. Certainly, small mammals appear, but not the top predators, large herbivores, nor birds that depended on old-growth forest.

The forests themselves did not do well. In the southeast, for example, long-leaf pine (*Pinus palustris*) was wide spread and abundant when the Spanish arrived in the sixteenth century. It made excellent ship's spars, and produced tar, pitch and turpentine, substances essential to the economy of the time. Although long-leaf pine is not extinct it now occurs naturally only in very small, disjunct areas (see: An Ecosystem Dependent on Fire pg. 57). In the natural course of recovery faster growing species have replaced it. Fire, once encouraged by native peoples or occurring naturally, is now replaced by occasional controlled burns. When allowed to proceed, fire cleared the understory, caused pine cones to open - thus spreading seeds, sterilized the upper layers of soil, and even affected soil pH by producing ash. These processes combined to maintain a healthy ecosystem.

Fragmentation of forests produces an edge effect, as there is a relative increase in the perimeter as forest area decreases. More edge around a forest patch can facilitate the predation of bird nests. As the forest interior becomes more accessible, avian nest parasites (i.e., brown-headed cowbird, *Molothrus ater*), and other predators including raccoons and opossums, gain access to ground- and tree-nesting bird species.

The absence of large predators such as mountain lion and wolf allows herbivores to flourish. Over-abundant deer browse heavily on shrubs, and young trees, which reduces both cover and the nest sites essential to birds. It now appears in some places that the pressure of rising deer populations threatens native plants such as ginseng (*Panax quinquefolius*). Ginseng is one of the most important medicinal plants harvested in the US. Further, pests and diseases weaken and reduce diversity in the forest. Other examples are the damage to the American elm (*Ulmus americana*) by Dutch elm disease and the devastation of the American chestnut (*Castanea dentata*) by the chestnut blight (*Cryphanectria parasitica*).

Worldwide, the disappearance of forests accelerated in the twentieth century and about half the world's forests have disappeared since 1900. The effects of human intervention shifted from the temperate forests to the tropics. While the temperate forest has stabilized in overall area, the tropical forests are now disappearing at unprecedented rates. The majority is cut not for lumber, but is cleared for agriculture. Forests that are logged for lumber generally grow back, but not always with the same species composition. Land cleared for agriculture is changed irrevocably, the soil quickly depleted or lost, a condition understood since the early 1600s. Conservationists point out that small farmers and growers are the leading cause of forest destruction, but global economic forces and short-sighted or corrupt government policies add weight and momentum to the destruction of the world's forests.

CATALPA

The southern catalpa or Indian bean tree provides the background for Catesby's portrait of the "Bastard Baltimore" or orchard oriole (*Icterus spurius*). His figure includes the flowers, leaves and seed-pod of this tree. Catesby claims to have introduced the tree to the inhabited parts of Carolina when he brought seed from inland Carolina. Apparently, locals instantly appreciated its arrival and Catesby reports that "…yet the uncommon Beauty of the Tree has induc'd them to propogate it; and 'tis become an Ornament to many of their Gardens…" He implies a similar success in England where he apparently introduced it around 1725. Cultivated for its attractiveness, it is now seen in suburban yards, public parks, and along the streets of London. Catesby noted the catalpa's most prized feature: "…it produces spreading Bunches of tubulous Flowers, like the common Fox-glove, white, only variegated with a few redish purple Spots and yellow Streaks on the Inside. The Calix is of a Copper-Colour" (I-49).

Catesby's oriole image served as Linnaeus's type for the species. Orchard orioles are common summer residents along the coastal plains of the east where they breed in forest groves and orchards. Insects provide up to 90% of the orchard oriole's diet, which they often supplement with fruits and berries, especially mulberries. At the times when catawba worms are active, orioles will feed heartily on them. Catawba worms also provide fodder for other birds such as red-eyed vireos (*Vireo olivaceus*). However, birds are not the only ones interested in the worms.

Catawba worms are the caterpillar stage of the catalpa sphinx moth (*Ceratomia catalpae*). The larvae are essentially black and yellow but color is variable and two phases may exist. They feed exclusively on catalpa leaves and can strip entire trees of their foliage. This is hardly desirable for ornamental shade trees. Larvae overwinter in the soil near previously infested trees and emerge after the trees have leafed out in the spring. Like many other insects, the timing of the emergence depends on temperature and general weather conditions. Infestations are sporadic and localized, and the insects can be abundant for up to three years.

Although recognized as a pest, the caterpillars are not without value. They are highly desirable bait for fish. Fishermen were known to collect them at least as early as the 1870s when entomologists first described the moth. A small industry exists now, and frozen worms are sold on the web and at bait and tackle stores in the South. In some places the more avid fishermen grow trees specifically for the worm harvest. When Catesby introduced the catalpa to coastal Carolina, he did a favor to more than just the horticulturists.

Carolina Jessamine
(*Gelsemium sempervirens*)

A native to South Carolina, jessamine is also called false jasmine, yellow jasmine and Carolina jasmine.

When grown in full sun it remains bush-like, but when shaded it will stretch to the sun and climb to 20 ft. on any available support including trellises, trees and buildings. The glossy, green leaves are lanceolate, 2-3 in. long and arranged in opposite pairs on the reddish stems. Small clusters of fragrant tubular yellow flowers are produced in the spring. The blossoms are about 2 in. long and have a flared lip much like honeysuckle. All parts of the plant can cause skin irritation.

It is found in temperate areas from Guatemala to the southeastern United States in pine and deciduous woodlands, on damp to very dry soils. An almost identical species occurs in wet woodlands and swamps, also in the southeastern United States; called the swamp jessamine (*Gelsemium rankinii*), its flowers are not fragrant.

I painted this specimen at a bed-and-breakfast in Summerville, South Carolina.

Gelsemium sempervirens — Summerville SC — Carolina Jassamine

Carolina Jessamine

Appropriately, the state flower of South Carolina, the Carolina jessamine was included in Catesby's plate illustrating the eastern phoebe (*Sayornis phoebe*). Carolina jessamine (Plate 7, pg. 55) is widespread in the wetlands of the southeastern states and is popular for landscaping in the region. It grows rapidly in full sun, and its fragrance can fill an entire yard. Jessamine climbs and spreads by twining, without tendrils, by which such familiar plants as gourds, peas, and grapes climb. Carolina jessamine is used commonly as a ground cover and can form an attractive shrub. Hardy to 15ºF, it remains semi-evergreen even in the coldest parts of its range.

Carolina jessamine is not generally browsed by deer, which is an added attraction for suburban gardeners. All parts of the plant contain poisonous chemicals. Jessamine also contains chemicals that, while not toxic, are recognized by the USDA for their notable physiological effects. In related plants (*G. elegans*, for instance), used in Chinese herbal medicines, the active agents are simple alkaloids. These drugs, derived from the root, can kill by arresting the center in the brain (medulla oblongata) that controls respiration. Not surprisingly, jessamine had ethnobotanical applications among the Native Americans of the pre-Columbian South. It is alleged to have relieved simple stomach ache, dysmenorrhea, gonorrhea, hysteria, coughs, short-windedness, and muscle spasms. Subsequent investigators have been unable to confirm that the plant is effective in the treatment of these ailments. Jessamine-derived drugs have also been implicated in homicide, presumably at much increased dosages. Nevertheless, jessamine has found a role in modern homeopathic medicine. It is used to treat arthritis, headaches, and flu. It has also been used at various times for nervous disorders, paralysis, spasmodic conditions (including asthma and whooping cough), stage fright, fever, and migraines.

An Ecosystem Dependent On Fire

A Forest and A Bird

When sixteenth century Spanish explorers arrived in the New World, long-leaf pine (*Pinus palustris*, Plate 8, pg. 59), was the dominant forest species from what is now Virginia to Florida and west to Texas. Much of the forest was nearly pure stands of the long-leaf, which arose from an understory that included grasses, herbs, and both sabal (or cabbage) palmetto (*Sabal palmetto*) and saw palmetto (*Serenoa repens*). Catesby notes that long-leaf pine barrens predominated on high sand hills between rivers and marshes. Fires spawned by summer thunderstorms and the occasional fire set by Native Americans kept the understory clear from competition from oaks and other hardwoods. Botanists estimate that the long-leaf pine savannah covered almost 144,000 square miles of the Southeast. Today, only about 3% remains.

Until the mid-1800s the forest was logged by hand. The invention of steam-powered sawmills and locomotives fostered industrial-scale logging. Besides the valuable wood, the trees were also a primary source of the resinous sap which provided tar, pitch and turpentine used to seal and protect wooden hulls.

For much of the nineteenth century feral hogs overran the Southeast, and fed heavily on long-leaf seedlings. Laws that required farmers to fence livestock came too late to provide protection. Fires became much less common due to active suppression by landowners. They were also passively prevented from spreading where roads and farm-fields acted as firebreaks.

Periodic burning was necessary to maintain the open environment of the long-leaf pine savannah. Ecologists estimate that fire occurred in some part of the primeval forest at least once every five years. As the understory was eliminated by these blazes, pines flourished. Long-leaf seedlings remained relatively short until swept by fire. Then a rapid growth spurt would follow, supported by the extensive root system that formed during the tree's first fire-free years. An exceptionally long tap-root provided additional support to the long-leaf. Of equal, or perhaps even more importance to the young trees' survival, was the protection provided to the terminal bud by the dense needle layer.

Human settlement patterns have interrupted the cycles that sustained the long-leaf forest. Farmers often burned the savannah's undergrowth in early spring to encourage fresh grasses for cattle. This was good for the cattle and farmers, but not for the trees. Annual burns reduced survival, as the seedlings did not have time to establish the root system so important to their rapid post-fire growth. It has taken foresters some time to realize that fire is not only critical to long-leaf survival, but that the periodicity of the fire was also crucial. Meanwhile the savannahs disappeared or underwent succession into denser hardwood forests.

One feature of the long-leaf pine is that it sheds its lower branches as it grows in height. This provided perfect sites for the nest holes of the red-cockaded woodpecker. This species

Red-Cockaded Woodpecker
(*Picoides borealis*)

Long-leaf Pine
(*Pinus palustris*)

Red-cockaded woodpeckers average 8 1/2 in. in length with a 14 1/2 in. wing span. Male and female plumages are similar with a black cap and neck streak, white cheek patch, and black and white barred horizontal streaks on back and wings.

A "cockade," is a ribbon or ornament worn on a hat. The "cockade" of the red-cockaded woodpecker is a tiny red line on the side of the head of the male. This is only visible during breeding season and difficult to spot. The juvenile males have a red spot on their black cap.

Their diet includes beetles, caterpillars, wood-boring insects, spiders and fruit when available. They do not migrate. In Catesby's time these woodpeckers ranged throughout the southeastern United States. Distribution today is restricted due to habitat loss.

The Long-leafed pine grows to a height of 80-100 ft. The species name, *palustris*, means, "of the marsh." Other names include longstraw, southern yellow, swamp, hard, pitch and Georgia pine. Its single trunk, covered in a scaly bark, can be up to 3 in. in diameter. The needle packet, in bundles of three, can grow to 18 in. long. These trees are slow growing and can take 200 years to reach full size. Their habitat is sandy acidic soils in full sun, from the coast to 2,300 ft. elevation. The small spring cones are followed by a larger female cone, 6-10 in. long. Historically the species ranged from Virginia to Florida, Louisiana to eastern Texas. Today it can be found sparingly in the temperate east of the United States as well as the southern states.

I painted this watercolor from preserved specimens of the red-cockaded woodpecker and long-leaf pine housed in the collections of the University of Connecticut.

Mountain Laurel
(Kalmia latifolia)

Sheep Laurel
(Kalmia angustifolia)

Bog Laurel
(Kalmia polifolia)

Catesby illustrated the mountain laurel in his second volume. He wrote:

> "After several unsuccessful Attempts to propagate it from Seeds, I procured Plants of it at several Times from *America*, but with little better Success; for they gradually diminished, and produced no Blossoms; 'till my curious Friend Mr. *Peter Collinson*, excited by a View of its dryed Specimens, and Description of it, procured some Plants of it from *Pensilvania*, which Climate being nearer to that of *England*, than from whence mine came, some Bunches of Blossoms were produced in *July* 1740, and in 1741, in my Garden at *Fulham*" (II-98).

The evergreen *Kalmia* species illustrated here are native to eastern North America. All parts of these shrubs are thought poisonous. The oval flowers of all *Kalmias* have ten pouches in which anthers are held under tension like bows until an insect releases them and consequently is covered in pollen.

Mountain laurel grows from 5-15 ft. with a gnarled trunk, and alternate, elliptical, 2-5 in. long, and shiny leathery, green leaves. Common names include spoonwood, as Native Americans used the wood to carve bowls and spoons. The flowers are in clusters, white and pink. Its range is from Maine to Florida and west to Louisiana and north to Illinois.

Sheep laurel has many names, sheepkill, lambkill, pig laurel and sheep poison. It is a mat-forming shrub growing 3 ft. tall and twice that in diameter. Its small size makes it accessible to grazing animals, thus the nicknames. The flowers are smaller than those of mountain laurel and are in clusters on the stem. Leaves are narrow and leathery, blue-green in color.

The diminutive bog laurel grows from 1-3 ft. tall with pink flowers at the end of stems and elliptical leaves that curl under at the edges, displaying white underneath. It grows from Labrador and Ontario south to Georgia, favoring wet, sphagnum bogs.

I was able to illustrate each of these plants from live specimens in Connecticut. Mountain Laurel is Connecticut's state flower.

(*Picoides borealis*, Plate 8, pg. 59) was historically found only in the open understory of the pine savannahs of the Southeast. As these forests disappeared, so did the potential nesting sites, and *P. borealis* is now listed as endangered. The species prefers to excavate nest holes in living, mature trees, below the lowest remaining branch, and in wood where red-heart fungus (*Phellinus pini*) occurs. The fungus appears frequently in older trees, but is apparently limited to the heartwood and is probably not fatal. Most other woodpeckers prefer to build nests in dead trees.

The red-cockaded woodpecker may take up to three years to excavate its nest hole, even in a diseased tree. The male is the primary excavator, but both sexes share the duties of incubation, brooding, feeding the young, and nest sanitation, although it is the male that incubates at night. Related species display a similar division of labor. However, unlike its relatives, the red-cockaded may live in extended family groups called clans and has an unusual breeding system. Early in nesting, the breeding pair may receive help in raising their brood from one or more, mostly male, offspring of the previous season. Breeding pairs with helpers tend to fledge more young than pairs without helpers. Helpers forego breeding for a season, but remain in the clan. The origin and possible advantage of such cooperative breeding is still debated. It generally occurs in species with relatively low rates of adult mortality. The helpers not only eventually gain access to a nest cavity which is a valuable resource, but also participate in the propagation of the family gene pool. Because the birds are territorial and non-migratory, a clan requires as much as 200 acres of old growth forest for foraging and nesting. Such conditions do not fit well with the depletion of the forest.

Another unusual relationship between bird and tree has to do with the red-cockaded's avoidance of predators. Because the nest cavity is located below the lowest remaining branch, predators cannot use branches to climb to the site, as a mammal might. But rat snakes (*Elaphe obsoleta*), a primary threat to the red-cockaded, can shinny up the trunk to gain access to the nest cavity. The woodpecker effectively defends the cavity by chipping small holes in the bark. These resin wells ooze sap down the tree trunk. Snakes cannot avoid contact with the resinous sap, which adheres to the scales on its underside. Even tiny amounts of the sap will inhibit movement, and prevent the snake from climbing further. Maintaining flow from the resin wells is time consuming and requires significant attention. Long-leafs produce considerably more sap than most trees, which is another reason they are sought out for nesting. If the tree damage from the wells becomes too great or the tree dies, the red-cockaded will abandon the tree.

The Laurels

Native to our eastern states, mountain laurel (Plate 9, pg. 61) grows as an understory shrub in many forests including long-leaf pine. This member of the heath family (Ericaceae) was first recorded in 1624. The name *Kalmia* honors Pehr Kalm, who collected specimens for Linnaeus. Kalm and Catesby met briefly in London in May 1748 the year before Catesby died.

In mixed woodlands where it is the principal understory plant, mountain laurel is capable of sending rhizomes 30 in. deep, where they are well protected against the heat of severe fire. While the oil and wax content of the leaves makes this plant highly flammable, mountain laurel can robustly regenerate from burls, layers, or suckers. Older-growth stands, without benefit of fire, form dense hedges that impede the establishment of pine seedling.

While culled for purposes of forest management, mountain laurel is a common garden plant and is one of our most cherished native shrubs. In the acid soil of the east, it is a long-lived plant with few pests. Although it is usually a shrub, in 1877, botanist Asa Gray noted at Caesar's Head in extreme northwest South Carolina that the trunks of mountain laurel reached 50 in. in circumference. Another common name, spoonwood, relates to the utensils Native Americans made from the wood. Because all parts of *Kalmia* are poisonous, it is not a favorite for foraging deer while the closely related sheep laurel (*Kalmia angustifolia*) has done in many a sheep and goat. Grayanotoxins are present in all *Kalmia* plant parts and ingestion can be lethal to pasture animals. These toxins occasionally turn up in both imported and local honey as they occur in rhododendrons and other members of the Ericaceae worldwide.

Tulip Tree:
A long drink of water

It is impossible to overstate the importance of plants to life on earth. Plants feed all of humanity and provide food and shelter for nearly all the earth's wildlife. Plants consume CO_2 and thus decrease greenhouse gases. Photosynthesis in plants produces the oxygen which, in turn, animals consume through respiration. Beyond this, plants enrich the soil by adding nutrients. The root systems can reduce erosion. Conservationists estimate that one third of all plants are threatened with extinction due largely to human activity.

Trees are defined as those plants with a permanent shoot system supporting a single woody trunk. Trees are familiar, successful, and as the most massive members of the plant kingdom, are among the largest living things on earth. Trees occur everywhere except the most extreme land environments. They take on a seemingly infinite variety of forms. It may appear wasteful to produce so much apparently unproductive tissues in the woody trunk, branches and roots, while only the leaves perform photosynthesis. But a permanent above-ground structure makes excellent sense. Because each plant must compete with others for water, nutrients, and especially light (the *photo-* in photosynthesis), the trunk, branches and roots are vital to procure the means to survive. This support system in deciduous plants doesn't have to be regenerated each year, but is in place when budding begins and the leaves appear. The diversity of trees is derived from differences in size and shape, differences in leaves, flowers and seeds, and even the density and color of the wood. The world's soil types and climates provide conditions for an astonishing combination of fine-tuned adaptations in trees.

Shrubs, by contrast, branch near the ground and typically have several slender stems rather than a single trunk. Stems can support many leaves but are less rigid than trunks and often bow down under their own weight. Understory shrubs often bud earlier and grow faster than trees, but are essentially limited to a shady existence overshadowed by their taller cousins. Trees grow more slowly, and live longer. This is due to the qualities of wood itself: it is strong and durable, and has a low need for maintenance, since most of its component heartwood cells are dead.

The tulip tree (*Liriodendron tulipifera*) or yellow poplar, is native to North America and was

Tulip Tree
(*Liriodendron tulipifera*)

Tulip trees range from New England west through southern Ontario and Michigan, south to Louisiana and east to central Florida. Common names include tulip tree, tulip poplar, yellow poplar, whitewood and canoe tree. The mature tree is tall (up to 190 ft.) and straight with deeply furrowed bark that creates a diamond-shape pattern. Leaves are alternate, 3-5 in. long and have a unique shape lacking a tip but with four to six lateral lobes. The 2 in. deep flowers are tulip-like with six petals, thus, the common name. Petals are yellowish-green at the tip and orange at the base.

Catesby commented:

> "THIS Tree grows to a very large Size; some of them being Thirty foot in Circumference....The Flowers have been always compared to Tulips; whence the Tree has received its Name; tho', I think, in Shape they resemble more the *Fritillaria*" (I-48).

I painted this tree from a specimen growing just across the street from us in Mystic, Connecticut. Over 25 years I have watched this tree grow at least 30 ft., obscuring the setting sun. Baltimore orioles build their nest high in the crown most every year.

introduced to Europe in 1637 by John Tradescant the Younger. It is not a poplar, but a member of the magnolia family.

Its flower resembles a large tulip (Plate 10, pg. 65), hence the common name. The yellow in yellow poplar derives from the color of the heartwood, which contains chemicals (terpenes and phenolics) that repel insects and fungi.

Tulip trees are the tallest trees in the eastern forest of North America. They grow to 80-100 ft., with occasional outliers like the "Sag Branch Giant" at 191 ft. 11 in., in the south-east reaches of the Great Smoky Mountains National Park at Boogerman Grove. The tree favors rich bottomland soil, and the wood is used in furniture making, valued for its long straight grain. Native Americans hollowed out large logs to make canoes. Tulip trees are prized ornamentals. The first leaves are enormous and extremely effective light-gatherers. With no change in shape, later leaves are smaller, adapting to the changing environmental conditions. In autumn, the leaves turn golden-yellow. The tulip tree produces abundant nectar that makes excellent honey.

Catesby portrayed the tulip tree with what he called the "Baltimore Bird". This was named after George Calvert (1579-1631), the first Lord Baltimore, because the males' bright orange and black plumage resembled Baltimore's banner. Catesby noted both its migratory habits and the unusual hanging nest. The "Baltimore Bird", now the Baltimore oriole (*Icterus galbula*), was for a short time called the northern oriole. Changes of this type in common names, especially in birds, reflect the struggle by ornithologists to agree on universal common names. Changes often appear capricious and only scientific names are recognized worldwide (see Organizing the Natural World, pg. 17).

Catesby remarks on the tulip tree and reported individuals with trunks 30 ft. in circumference. Large examples have lived for 150 years. Just for scale, the massive giant redwoods (*Sequoiadendron giganteum*) grow to over 300 ft., twice the height of Pisa's Leaning Tower. Extreme height interests plant physiologists because it presents the problem of transport: what is the mechanism by which the tree lifts water absorbed through the roots up to the leaves at the crown?

Efficient water transport can enhance carbon dioxide consumption and improves both growth and reproduction. Straightforward mechanisms such as osmotic pressure and capillary action have been suggested and disproved. The force of drawing the liquid up, as with a straw, is inadequate to lift water to this height. In fact, the water is pulled up under tension as it replaces that lost through the leaves by transpiration. Water in an extremely narrow channel can withstand large stretching forces without breaking into smaller subunits. This strength is derived from the cohesion between the water molecules. As it is pulled higher, water is actually being stretched in the tree! The wood structure of angiosperms, such as the tulip tree, is a highly efficient water-transporter. The narrow transporting vessels are produced in the new wood grown each spring, added on as growth rings to the denser mechanical trunk structure. In maple trees, the sap is harvested in early spring, reduced by boiling, and sold as maple syrup.

Whatever the mechanics, the ability of trees to conduct water declines with age. The older wood, known as heartwood, fills with gums and resins which repel disease and wood-boring insects but cannot conduct water. Heartwood is very strong, stiff and provides great strength and support.

The sapwood, by contrast, is softer and continues to conduct the life-bearing liquid. Though trees can live to great old age, this structural system is prone to failure in dry or freezing weather. Hence trees are vulnerable to drought. Dry conditions can lead to the death of the leaves, which usually begins at the tree's top and proceeds downward. In cold conditions, transpiration ceases, the leaves are shed, and dormancy ensues.

Water transport may also limit how high trees can grow. The limit is not in the leaf bud, as experiments have shown that vigorous growth reappears when buds from old, tall trees are grafted onto younger trees. Trees seem to stop growing when the water delivered to their leaves become insufficient for photosynthesis. This does not mean that they die, only that they cease growing upwards. How this physiological difference between young and old trees stops growth is not presently understood.

OAKS

A few plants native to the Carolinas would have been known to Catesby before his arrival on these shores. Early colonists, especially from Virginia, had sent seeds to England, and so Americaan species appeared in gardens in and around London. As early as 1617, John Tradescant the Elder (1570-1638) had purchased American cowslip (*Primula meadia*), and false black locust (*Robinia pseudoacacia*) for his garden. His son John, the Younger,(1608-1662) visited Virginia in 1637, 1642, and 1654, and returned with bald cypress (*Taxodium ascendens*), tulip trees (*Liriodendron tulipifera*), red maple (*Acer rubrum*), Eastern black walnut (*Juglans nigra*), red mulberry (*Morus rubra*), shagbark hickory (*Carya ovata*), Virginia creeper (*Parthenocissus quinquefolia*), American sycamore (*Platanus occidentalis*) and even poison ivy (*Toxicodendron radicans*). Native to Virginia, most of these species prospered under cultivation, and, fortunately, poison ivy did not become invasive in England. Other contemporary collectors also returned with North American native coneflower (*Rudbeckia laciniata*), goldenrod (*Solidago canadensis*) and the rain-lily (*Zephyranthes atamasca*).

According to Catesby, Carolina had three types of cultivated soils "…distinguished by the Names of *Rice Land, Oak* and *Hiccory Land*, and *Pine barren Land*." Each had come under the influence of man:

> "*Rice Land* is most valuable, though only productive of that Grain, it being too wet for any thing else. The Scituation of this Land is various, but always low, and usually at the Head of Creeks and Rivers, and before they are cleared of Wood are called *Swamps*. . . . These Swamps, before they are prepared for Rice, are thick, over-grown with Underwood and lofty Trees of mighty Bulk, which by excluding the Sun's Beams, and preventing the Exhalation of these stagnating Waters, occasions the Land to be always wet, but by cutting down the Wood is partly evaporated, and the Earth better adapted to the Culture of Rice; yet great Rains, which usually fall at the latter Part of the Summer, raises the Water two or three Feet, and frequently cover the Rice wholly, which nevertheless, though it usually remains in that State for some Weeks, receives no Detriment" (I-iii).

WILD OLIVE
(*Osmanthus americanus*)

Catesby called this tree "The Purple-Berried Bay". He wrote:

> "THIS Tree grows usually Sixteen Feet high: And the Trunc is from six to eight Inches in Diameter. The Leaves are very smooth, and of a brighter Green than the common Bay-Tree: Otherwise, in Shape and Manner of Growing, it resembles it. In *March*, from between the Leaves, shoot forth Spikes, two or three Inches in Length, consisting of tetrapetalous very small white Flowers, growing opposite to each other, on Foot-stalks half an Inch long. The Fruit, which succeeds, are globular Berries, about the Size of Those of the Bay, and cover'd with a Purple colour'd Skin, enclosing a Kernel, which divides in the Middle" (I-61).

Nicknamed devilwood or tea olive, this is our only native olive. Other *Osmanthus* species are native to Asia. It is an evergreen tree or shrub that grows 15-25 ft. and occasionally 30-40 ft. in the wild. The common name devilwood comes from the difficulty of trying to work its dense, gnarled wood. Its distribution is south coastal Virginia to central Florida west to south east Louisiana.

The springtime clusters of small, creamy, tubular, fragrant flowers are followed by blue-black fruits. Leaves are opposite, shiny, green, elongated, 2-4 in. long, with a smooth edge. *Osmanthus* has few pests or diseases. It grows well along swamp margins and streams, tolerating both flooding and salt spray.

This watercolor was painted from a tree on the grounds of our bed-and-breakfast in Summerville, South Carolina in the early spring.

Osmanthus fragrans   Tea Olive

This is both a grand description of the local conditions and documentation of the changes engendered by the human presence (see Rice, pg. 40). Catesby reports:

> "The next Land in Esteem is that called *Oak* and *Hiccory-Land*; those Trees, particularly the latter, being observed to grow mostly on good Land. This Land is of most Use, in general producing the best Grain, Pulse, Roots, and Herbage, and is not liable to Inundations;..." (1-iv) The trees, of course, are excellent for lumber (especially oak) and fuel (especially hickory). "The third and worst Kind of Land is the *Pine barren Land*, the Name implying its Character." While the soil is poor, the trees yield "...beneficial Commodities, of absolute Use in Shiping, and other Uses, such as Masts, Timber, &c. Pitch, Tar, Rosin and Turpentine." (I-iv)

This picture of land usage pertains to the settled lands of the colony. The unsettled portions were described as tracts that included "*Bay-Swamps*," typically permanent wetlands intermixed with pine stands. The plants here included aquatic shrubs such as "*Red Bay, Water-Tupelo, Alaternus, Whorts, Smilax,*" and others. Catesby found these tracts profuse with flowers and accommodating to waterfowl and wading birds. He also mentions, but apparently did not visit "*Scrubby Oak-Land,...more sterril than that of Pine barren Land*"(I-iv) and not at all adaptable to cultivation.

Catesby's collections included nine species of oak (*Quercus*). *Quercus* is a large genus of about 500 species, distributed throughout the North Temperate zone. Birds, swine, and squirrels, as well as by pre-Columbian Indians ate the oak's acorns. The tannins of the oak are natural dyes; the bark of some provides cork, and excellent wood for fuel, the construction of homes, furniture, boats, and the barrels used to age whisky and Chardonnay.

The live oak, *Q. virginiana*, very common in the Southeast, appears in Catesby's plate of the pileated woodpecker, *Dryocopus pileatus*. He reports the acorns to be among the sweetest and used by the Indians to thicken venison soup.

Oaks are valued not only for their uses, but for their aesthetic appeal. The oak leaf has been used as a symbol of strength and courage. It appears in decorative carvings on furniture and fireplace mantles. In the military, the oak leaf can denote additional honors upon a particular decoration, as in a Bronze Star with Oak Leaf Clusters. State flags, the logos of countless financial institutions, and various quasi-official signs and banners include the oak leaf for its symbolic connotations.

WILD OLIVE

Throughout the coastal south and on barrier islands, live oaks are often found with wild olive, *Osmanthus americanus*. Wild olive, or devilwood (Plate 11, pg. 69), a member of the olive family (*Oleaceae*), is a large evergreen shrub that prefers dappled shade and moist conditions. As it is tolerant of salt spray it does well in seaside gardens. It is noted for an extremely fragrant flowers said to almost put the gardenia to shame.

About Leaves

Leaf shape has an interest beyond simple symbolism. The leaf is the interface between plants and their environment. Leaves are generally flat which maximizes their exposure to the sun and similar in their internal workings (e.g., chemistry and physiology). However, their silhouettes vary widely, and a huge number of shapes have been described.

Leaves are one of nature's treasures. Metabolically, leaves are the organ that produces the basic food, which is the fuel, for plant function. Energy from sunlight is harvested in photosynthesis and used in the plants' growth and maintenance. These are processes crucial to the life of all green plants. With the solar energy, photosynthesis converts carbon dioxide from the air into sugars and complex polysaccharides. The green color of leaves is due to the presence of chlorophyll, the pigment that is key to the process. Once produced, the polysaccharides are eventually metabolized providing energy to the plant, during which oxygen is released. In turn, atmospheric oxygen plays a critical role in animal respiration.

When organic materials combust or decompose, both energy and carbon dioxide are released into the atmosphere, completing one of the most important cycles supporting life on earth. Vast stores of organic materials, the remains of plants or animals, exist in the ground as peat, coal and oil. The processes of photosynthesis and plant respiration have remained the same for at least the last one billion years, and we owe to the chlorophyll of ancient plants the fossil fuels we burn today. Leaves may also provide food for other organisms as the polysaccharides make their way into the food chain. Leaf-eating, browsing herbivores are obvious beneficiaries. Leaves also provide a complete habitat for leaf-chomping, burrowing insects.

Functionally, the carbon dioxide used in photosynthesis is absorbed through small pores–the stomata–through which water is lost by evaporation. Water is absorbed by plants through the root system, an extensive, underground structure for nutrient collection and support. The leaf that sits at the atmospheric interface balances the requirements for light, heat, moisture and carbon dioxide. The size of the absorptive surface and the presence of lobes and complex edges are important attributes and can change with their position in the canopy. They reflect the availability of water and nutrients. Any stand of trees or forest with multiple layers will have plants with leaves of different shapes, but all with similar functions.

Shape and area are related; compare for example, a beech tree with a spruce or pine needle. As leaves ultimately determine the energy budget for the season, trees with smaller total leaf area, such as pine and spruce trees, have lower rates of photosynthesis, though they are green year round and can work more days per year. An example of extreme reduction in leaf area is the palo verde tree (*Parkinsonia* spp.). Found in hot dry areas of the southwest, its leaf area is absolutely minimized, which reduces water loss through evaporation. But the palo verde has green bark on the trunk and branches, where photosynthesis takes place. Plants are also configured in their various layers and levels so that leaves do not overlap, and therefore shade each other out.

Leaf shape is defined by edges in space. And shape – linear, lanceolate, spatulate, palmate, whatever – is a consequence of the pattern of growth. Leaves, no matter what their size and shape, must satisfy the simultaneous requirements for mechanical stability, water conservation, and light harvesting. Consequently, there is no optimal shape or size for leaves. The fierce competition for light within the complexity of the plant community generates much effective morphology. The simply shaped willow leaf at the stream's edge is no worse or better at doing its job than the huge split-leaf philodendron of the rain forest, or the oak or maple.

Thugs and Aliens

Observing a community of native plants is a bit like looking into a kaleidoscope where the pattern of color, size, and shape is the result of composition, color and geometry of the elements. History has an effect: turn the tube and a new pattern will emerge. The same units are cast into a new configuration and consequently a different pattern emerges. Change the shape or size of one or more elements or add a new element and yet another, albeit related, pattern emerges. Patterns never precisely repeat as history unfolds. Shake or roll the tube to appreciate further the often unpredictable interactions in nature.

In his collecting New World specimens, Mark Catesby anticipated the story of wholesale movement of alien species around the globe—one of the least predictable series of interactions the natural world has faced for millennia. With the arrival of any invasive organism, the mix of species is changed, even when the community appears stable. The compositional plasticity of the native flora community is challenged. The question arises: how much internal change can be tolerated while a general equilibrium in the system is maintained? Will a successful coexistence evolve or will the alien overwhelm a balanced community? Predicting the future of the system will depend on knowing the past and the properties of the alien. Especially now, as alien species proliferate in many places, the calculus of prediction is complicated when the arrival of one invader works on subsequent arrivals in the network.

Indigenous or native floras are the product of an incompletely known history. For example, the continued survival of an endemic plant community depends on an array of environmental conditions that include soil composition, rain, temperature, insolation (the total energy received from solar radiation), fire, and other disturbances, all of which vary seasonally and may undergo long-term cyclic variation. In addition to these physical factors, biotic factors – the endemic insect fauna of pollinators and herbivores, parasites, and diseases, for instance – also play roles in determining the long-term success of given species. The genetic variability inherent in each species is equally important in determining its ability to respond to environmental pressures through selection.

Undoubtedly, species composition in a given habitat has always varied over time. In any community, species makeup is determined largely by the nature of those species that are present after a natural disturbance, such as climate change or changes in available resources. Over time, a community may appear stable superficially while both individual species and the overall composition may evolve. Communities in similar habitats but on different continents may share similar functions and component parts, but consist of entirely different species.

Communities are always open to invasion. One potential vector is the continual influx of plant seeds. The dispersal mechanisms for plants are almost as diverse as the plants themselves. Besides physical forces such as the wind, animals are also agents of seed dispersal. Birds, for example, eat fruits such as the berry of mistletoes (Loranthaceae) and disperse seeds through their guts. Similarly, large grazing or browsing mammals facilitate seed dispersal, as do insects of many types. However, only comparatively few plants become established in any particular community.

In recent times, human activities have become an overwhelming force in species dispersal. Vastly increased movements of people and goods have globalized many species and diseases both intentionally and unintentionally. Alien seeds and plants are usually out-competed by the natives. What is most notable is the occasional alien that becomes outrageously successful. What makes a plant invasive and another not, depends on myriad factors. An invader may prevail due to greater reproductive capacities, better dispersal abilities, or particular growth patterns. Perhaps it escapes a predator, parasite, or disease that previously controlled it. With such controls gone, the species can run rampant. The introduction of plants, animals, and microbes worldwide often has impacts on economics as well as on biodiversity.

Native flora securely embedded in a diverse network of interacting elements is more stable and exists for longer periods than a flora in transition. While one species may replace a similar one with little or no change in the system's overall properties, this flexibility in composition makes the community more, not less, resistant to dramatic change. An alien invasion, perhaps coupled with other anthropogenic disturbances, could trigger a cycle of changes with an entirely unpredictable outcome. Because the composition of any community at any particular time is essentially the result of a lottery, it is also possible that the organization of the community itself contributes to the success of an invasive species. Some of the indigenous communities Catesby encountered may have survived apparently unchanged until the present, but it is almost certain that changes, independent of the work of man, have happened. At the same time, the plants he sent to England were aliens there, but very few became invasive. The arrival of non-native species need not be a disaster. Every community of plants and animals is in a state of dynamic equilibrium dependent on the arrival of new forms and replacement of the old. If a native species is out-competed it may become locally extinct, and the health of the community might be unimpaired. Competition is necessary to maintain vitality, but it is not possible to predict the details of the emergent community.

Internal ecological changes imposed by the alien, such as the availability of light or water, contribute to the process of restoration. There is no guarantee that a community will be any more successful at one time, with one set of species, than any other; however, if diversity is related to stability, then the least stable condition is a monoculture. And monoculture communities have frequently accompanied human settlement. Simple, single species systems are often vulnerable to a single weather event such as a drought, as occurred during the American dust bowl of the 1930s, or the outbreak of a disease such as the potato blight which caused famine in Ireland in the late 1840s. Events like these can cause economic turbulence and great human displacement.

Biologists have made progress in identifying traits of alien organisms that may predict their success in a new environment. In North America introduced mammals have included the horse, burro, pig, cattle, and goats. Spectacularly successful alien birds include the house sparrow

(*Passer domesticus*), European starling (*Sturnus vulgaris*), and the rock pigeon (*Columba livia*). The ring-neck pheasant (*Phasianus colchicus*), has also established itself well in its new American habitat. Recently, there have been serious outbreaks of exotic insects: gypsy moth (*Lymantria dispar*) and hemlock woolly adelgid (*Adelges piceae*). Aggressively spreading introduced plants include purple loosestrife (*Lythrum salicaria*), Kentucky bluegrass (*Poa pratensis*), several knapweeds (*Centaurea* spp.), numerous eucalypts, and kudzu (*Pueraria montana*), now seemingly ubiquitous in the Southeast.

Many species, especially Australian pine (*Casuarina equisetifolia*), water hyacinth (*Eichhornia crassipes*), Brazilian pepper-tree (*Schinus terebinthifolius*, Plate 13, pg. 81), and oriental bittersweet (*Celastrus orbiculatus*) are widespread ecological disasters. Successfully introduced microbes include those causing chestnut blight, smallpox, anthrax, hoof and mouth disease, and Dutch elm disease (*Ophiostoma ulmia*). On a geological time scale alien species may be short lived and transient. However, while they may leave no fossil record, their ecological influence can be significant.

One recent example is the zebra mussel (*Dreissena polymorpha*). Introduced in the Great Lakes, probably in discharged ship ballast, zebra mussels are filter feeders rather than grazers. Their free-living larvae facilitated the species' rapid spread, and they easily form vigorous colonies in new locations, including the water intakes of power plants. This fouling is a costly consequence of the mussel's invasion. As a result of their feeding behavior, industrial pollutants, such as dissolved PCB and heavy metals, quickly accumulate in their tissues, and subsequently enter the food chain. The effect on waterfowl that eat the mussels has been dramatic. Populations of mussel-eating greater (*Aythya marila*) and lesser (*A. affinis*) scaup, and bufflehead (*Bucephala albeola*) became contaminated and their numbers plunged. Related waterfowl with different feeding habits such as the common goldeneye (*B. clangula*), long-tailed duck (*Clangula hyemalis*) and white-winged scoter (*Melanitta fusca*), remain unaffected and even increased in some places. In many areas, zebra mussels have out-competed and excluded native freshwater mussels. In time, the zebra mussel itself became a resource as prey for several fish species that fed heavily on them. As the mussels decreased, exuberant communities of freshwater sponges developed and trematode–parasitic flatworms–outbreaks increased. Both of these have harmed mussel growth and survival. In the unpredictable events that follow the appearance of invasive aliens, the interactions operate in many ways.

One possible basis for the success of alien organisms is freedom from the pathogens in their native environment. For example, some plant invaders from Europe suffer fewer viral and fungal infections in North America than native species. Invasive animals, too, may be healthier. The European green crab (*Carcinus maenus*) was introduced along the New England coast in the late 1800s with rock ballast dumped from ships. Along their native European and North African Atlantic coasts, the crabs are small, rare, and plagued with parasites that stunt growth and inhibit reproduction. In the United States the crabs are large, common, and relatively free of parasites. They have become a dominant predator and feed heavily enough on local shellfish to do significant economic harm. The success of introduced species has been described as the process of ecological release. The new environment is more salubrious, and when competition, predation and natural diseases are left behind, an alien can vigorously breed and spread. In theory, unwanted invaders could be controlled by introduction of a disease, predator or parasites from their native habitat. Possibly, this would restore a state of equilibrium. This strategy is not without precedence; nor, of course, is it without risk. An introduced herbivore or parasite can ignore the original host and find (and destroy) a native species. Invasive organisms are second after habitat destruction as a threat to biodiversity worldwide.

Invaders can also drive the evolution of a native species. To demonstrate an actual evolutionary change requires that the native population change genetically as a result of the interaction with exotic species. Some of the best examples occur among phytophagous (plant-eating) insects shifting to a new host, the introduced species. Native insects have also adapted to introduced host plants that were initially toxic. As invasive species interact with natives, novel co-evolutionary relationships, direct and indirect, may appear. The ultimate impact of the introduction will depend on some combination of the traits of both the invader and native species in the community. Ultimately, the ecological effects can ramify through the entire web.

Prior to 1800, explorers and collectors introduced about 110 species of trees to Britain. An additional 200 species arrived between 1800 and 1900. Few of the tree species introduced in the twentieth century have yet become naturalized. Flowering plants, other than trees, have been considerably more successful and are perhaps more dangerous, especially as a threat to the native flora putting the biodiversity of large areas at risk.

Empress Tree

Although native to China, the empress tree (Plate 12, pg. 79) is named after Anna Paulowna (1795-1865), daughter of Tsar Paul I of Russia. Anna married William of Orange who became William II, King of the Netherlands. Anna was Queen Consort from 1840 to her death in 1865. According to The Philadelphia Historic Plants Consortium (http://growinghistory.worldpress.com), Philip Franz von Siebold, a physician from the south of Germany, worked for the Dutch military in Japan the 1830s. Siebold sent plants and seeds to Europe and one was *Paulownia* that was sent to France. Daniel J. Browne, in his 1846 *Trees of America*, noted that the *Paulownia* was in the Jardin des Plantes in Paris. He also wrote that *Paulownia* was introduced to the United States via Parson's Nursery located in Queens, New York City in 1843, but it most likely also came via other avenues as well. While primarily ornamental in Europe and America, it had been used for medicinal purposes and for timber in China as early as the third century BC.

The species epithet, *tomentosa*, refers to the hairy surface of the leaf. Hair-like structures are common on plants and are made of cellulose, a polysaccharide (unlike mammalian hair, made of the protein keratin). The leaf's shape is similar to that of *Catalpa*, but the two species are not related. Princess tree (an alternative common name) is a member of the Scrophulariaceae (the figworts), or is placed in its own family, Paulowniaceae.

The empress tree grows fast and has prodigious seed production. Its rapid growth is mythologized in China where tradition was that when a girl was born a tree was planted. The tree grows along with the child. When she becomes eligible to marry, the then mature tree is harvested and carved into wooden articles for her dowry. The wood today is still made into kotos, the stringed zither-like instrument of Japan and Korea. Woodworkers in the United States use it in furniture. In the nineteenth century the seeds were used as packing material for porcelain being shipped from China, in much the same way polystyrene "peanuts" are used today. As a single tree will produce 20 million seeds annually, they were an abundant resource. Both strong and fluffy, the seeds provided excellent protection for the delicate Chinese exports. However, when the packages leaked or broke open in transit, the seeds were scattered along railroad tracks or roads. Where soil and climate were conducive, the species became invasive. This was especially true along the eastern seaboard of the United States and in Japan.

Because the seeds are easily carried by wind and water, *Paulownia* spreads readily. The flowers can be pollinated by a variety of insects, adding to its success, as does an extensive root system, which helps it survive fire. The empress tree is a popular nursery plant, but the heavy seed production is considered a disadvantage in landscape gardening, as it becomes a potential menace to neighbors who may not want volunteers sprouting on their property. Consequently, nurserymen have produced cloned trees that are essentially sterile. There are no seed-pods to drop and hence invasiveness is greatly reduced. The showy, pale violet, trumpet-shaped flowers persist and decorate the landscape each spring.

Black Locust

The black locust (*Robinia pseudoacacia*) is native to the southern Appalachian region, but was not recorded by Catesby. It can spread aggressively and moves quickly into disturbed areas such as roadways, abandoned fields and strip-mined hillsides. A legume, it spreads initially by seeds, but soon builds a thick system of root suckers. In pure stands, it out-competes other woody plants and even weeds. Since all the trees in a stand originated vegetatively they share the same genotype and a large cloned organism is produced. Two features make the wood valuable to humans. First, the trees grow quickly, reaching sufficient size to harvest in relatively few years. Second, the olive-green wood has natural anti-rotting properties. The logs make excellent fence posts, railroad ties, and first-rate fuel. The extensive root system also controls erosion. Clearly, it is a very useful native species.

While black locust is naturally invasive, its spread through the east and upper Midwest has depended on human translocation. Because it spreads locally through root suckers, new roots and shoots interconnect to form dense, fibrous groves. Spread is actually encouraged by root damage, as it stimulates vigorous sprouting. Cutting, disease, wind, and fire are common sources of damage. The black locust is also harmed by locust borers, locust leaf miners, and twig borers. Pathogens such as bark canker and tree rot also afflict the black locust. All these encourage root spread and colonies eventually persist as a group of fairly sizable trees. This tree is a survivor, which is a characteristic of a successful invasive species.

As mentioned, the trees are useful, but difficult to control or eradicate. Cutting, mowing and burning all promote growth by sprouting or seed germination. Chemical control is difficult because of the wide-spread root system. In effect, the black locust's range expansion is an example of an invading native species. While black locust can be detrimental to native shrubs, it is not considered as threatening as many alien species. However, it does contribute to ecological homogenization and decreases biodiversity locally.

Brazilian Pepper-tree

Alien plants, among the most aggressive invasives, have a long history of environmental damage. Despite many local successes no single plant has expanded to become dominant worldwide. Both biological and non-biological constraints prevent total conquest. Nevertheless, powerful as these factors are, occasionally a plant species becomes aggressively invasive and rapid expansion will follow. In the southeast, water hyacinth (*Eichhornia crassipes*) is abundant enough in waterways, lakes, and lagoons to restrict navigation and reduce water abatement in flood control channels. Some think that the Brazilian pepper-tree (Plate 13, pg. 81) is on its way to displace every plant or tree in Florida.

The uncontrolled expansion of the Brazilian pepper-tree is a serious ecological threat to Florida's natural eco-systems. Introduced into North America about a century after Catesby as an ornamental, it has become the most widespread exotic in Florida. Spread by birds that eat the fruits, it pioneers in disturbed areas and spreads into undisturbed natural areas as well. Its vigorous spread is based on its ability to commandeer a community's resources. The number of plants may not moderate until a predator, disease, or a stronger competitor such as a human with a chain saw, backhoe, and herbicides arrives.

Aside from invasiveness, Brazilian pepper-tree is also a health menace. It is a relative of poison ivy (*Toxicodendron radicans*), and exposure to the sap can cause severe and persistent skin irritation. The crushed leaves, which smell like turpentine, release airborne chemicals that cause nasal congestion, sneezing and eye irritation in some individuals. Eating the leaves can cause hemorrhages and fatal internal conditions in horses and cattle. Birds may consume berries in quantities that cause intoxication and death.

Brazilian pepper-tree, along with kudzu (*Pueraria lobata*) and Australian pine (*Casuarina equisetifolia*), are costly and damaging aliens to the south east United States. Control or eradication is expensive. One avenue is a search for an appropriate biological control agent that works directly on the plant, such as a specific disease, or one that interferes with the pollinating insects. State laws that prohibit sale, cultivation, or transport may be difficult to enforce. Management techniques may include mechanical removal and widespread application of herbicides, but there may be unexpected side effects.

EMPRESS TREE
(*Paulownia tomentosa*)

This *Paulownia* species has naturalized in the eastern United States from Massachusetts to Texas and is found by streams and roadsides. Common names are the empress tree, princess tree and foxglove tree.

It is a fast growing deciduous tree, and reaches 60 ft. in height. The bark is smooth, gray-brown. Leaves are ovate and opposite, some 12 in. long with five lobes. Both sides of leaves are hirsute. Flowers are fragrant and funnel-shaped, pink to lavender resembling a foxglove (*Digitalis purpurea*) flower, and appear in bunches before the leaves. The seeds are contained in a tan, oval capsule 1 1/2 in. long. This tree now has the reputation as an exotic invasive in North America although it has been awarded the Royal Horticulture Society's Award of Garden Merit in 1993.

In Catesby's time, this tree had not yet begun its journey to North America, nor Western Europe. I cannot help but think how much he would have been taken with it, given his great interest in the cultivation of our native *Catalpa* in English gardens.

I have painted this tree from several sketches done in North Carolina and coastal locations north to Connecticut.

Brazilian Pepper-tree
(*Schinus terebinthifolius*)

Brazilian pepper-tree is a shrub or small tree that can reach 30 ft. or higher. It has a short trunk with many branches, some procumbent, some trailing or climbing. Common names are Brazilian pepper, Florida holly, Christmas berry and pepper tree.

The compound leaves are 1-2 in. long and have 5-9 in. elliptical or slightly elongated leaflets, arranged in pairs with a single leaflet at the tip. The leaflets are shiny and have slightly toothed margins. Small white flowers occur in clusters from the leaf axils of stems. The male and female flowers, occurring on different plants, have five small whitish-coloured petals. The small fruits are 1/4 in. in diameter, glossy and bright red when ripe. They possess a thin outer skin, soft pulpy middle, and hard stony central part that encloses a seed. The plant is cold sensitive and distribution is limited to the warmer areas of southern and central Florida north to the Georgia border.

I had no problem finding *Schinus* in Florida. It blankets areas totally, and the fruit is a favorite food of birds.

Morrow's Honeysuckle and Cedar Waxwing

The honeysuckles are a group of erect or climbing shrubs or vines with generally showy and fragrant flowers. In fruiting, they set colorful, small berries that ripen in late summer. Morrow's honeysuckle (*L. morrowii*) is native to Japan, and was described by Asa Gray from specimens collected by S. Wells Williams and James Morrow during the expedition of an American squadron to the China seas and Japan 1852-1854, under the command of Commodore M. C. Perry. Gray was one of Darwin's strongest advocates. The honeysuckles are planted as urban ornamentals and in semi-rural areas for wildlife habitat. They provide both cover and food, which makes them popular with gardeners who wish to attract birds. The bushes form dense thickets bordering woods and shores; its hardy nature, tolerance of drought, and ability to spread, especially along highways, make this honeysuckle attractive for low-maintenance plantings.

The dark side of these alien plants is that they are highly invasive. They spread aggressively and do very well in disturbed habitats. However, their prolificacy has caused them to be "barred" in Connecticut, "prohibited" in New Hampshire, and considered a "class B noxious weed" in Vermont and about 40 other states. Nevertheless, Morrow's honeysuckle is now widespread in the Northeast. The Morrow's honeysuckle's invasiveness and capacity for displacing native species has not been its only legacy. There are those who love it. Cedar waxwings (*Bombycilla cedrorum*, Plate 14, pg. 85) are nomadic birds that thrive on a diet of sugary fruits. They often breed later in the summer than other songbird species. Catesby named it "The Chatterer" presumably from its simple song which consists of a series of irregular, high screee notes. Among birds in a small flock in a stand of cedars, this often sounds like aimless chatter. The secondary feathers of the wing often bear a red, wax-like tip providing the common name. It is the only North American bird with yellow on the tips of the tail feathers, which produces a striking banded pattern.

In the 1980s, researchers in western Pennsylvania were amazed when young cedar waxwings they were monitoring turned-up with red, rather than yellow bands on their tails. Chemical analysis showed that the pigment, a carotenoid, was not the same one present in the yellow and reds of the adult plumage. Carotenoids are not produced by animals but derived from the diet (see Ibises: Red and White, pg. 89). While the food-borne molecules may be modified before they are deposited in the feather, this is not always the case. Either way, deposition occurs only while the feather is growing. Knowing this and knowing that recently hatched birds hold the juvenile plumage into early autumn, it followed that the cedar waxwing's diet as the feather grew would be responsible for the unusual color observed by the researchers. The red pigment in these individuals was rhodoxanthin, not the yellow xanthophylls normally present. When berry-producing plants in the area were surveyed, only one, *L. morrowii*, contained molecules that matched those in the tail feathers. The berries yielded rhodoxanthin which is present in many other plant species and which is deposited directly in the feathers, inducing the color change.

This is a tale of unintended consequences, but one that is seemingly benign. The change in plumage color of the cedar waxwing is related to the introduction of a particular plant species and appears only in areas where the plant is found. Careful observation has subsequently shown that individual young-of-the-year cedar waxwings generally have orange or red tipped tails in late summer or early fall. This color is produced by the diet during the previous molt. The normal plumage color returns in the following breeding season; hence there is no selection on this transient variation. The diet consumed the following spring contains adequate precursor pigments to produce the typical yellow feathers.

Purple Loosestrife

Florida has the most species of plants, the greatest biodiversity, and the greatest number of invasives of any state or province in North America. There are about 4,000 plant species recorded in Florida and 1,200 are non-indigenous. Some of the latter are serious invasives. Ironically, fully 90% of the non-natives were introduced deliberately. People, wherever they live, seem to like exotic plants, especially fast-growing trees. Also, it turns out, humans are drawn to exotic animals as pets and game animals. However good, or at least harmless, the intention, these proclivities have brought in all manner of plants and animals which are currently set on an exceptionally complex collision course with native species. The situation is made more threatening because people also seem to favor disturbed or heavily managed environments such as golf courses, neat lawns defined by well-tended flower beds, weed-free fields and orderly woodlands. Invasive species in such landscapes can have huge and lasting effects.

Consider purple loosestrife (*Lythrum salicaria*), which was introduced to the northeastern United States and Canada in the nineteenth century for ornamental and medicinal use. It is a perennial herb that, depending on conditions, can grow from 4-10 ft. tall and produces an attractive magenta-colored flower spike. With high fecundity and good seed dispersal, purple loosestrife readily adapts to both natural and disturbed habitats where it quickly replaces native grasses, sedges and other flowering plants. It has an extended flowering season that results in the production of enormous quantities of seed, and it also reproduces asexually through underground stems. Purple loosestrife's dense homogeneous stands restrict native plants, including endangered orchid species, and reduces waterfowl habitats. The self-replacing stands along lake and stream margins and practically all wetlands appear permanent. It has become a severe ecological threat.

Purple loosestrife is a textbook example of a successful alien: a thug. Understanding how such an invasion happens may lead to insights into how to control or exterminate the invader. Presumably the plant has some advantage, however small, over the native plants it replaces, either in its ability to disperse, or by being free of some natural mechanism that controlled it in its native habitat. Under experimental conditions, purple loosestrife from North America will grow taller and produce greater biomass than plants from its native Europe. This suggests that North American plants may have escaped one or another of factors that limit growth. If, for example, the new environment changes and a particular herbivore (one that may have specialized

Morrow's Honeysuckle
(*Lonicera morrowii*)

Cedar Waxwing
(*Bombycilla cedrorum*)

The cedar waxwing is a medium-sized bird with a brown head and chest. There is a noticeable crest atop its head. It has pale brown-gray wings, a pale yellow belly and a short gray tail with a bright yellow tip. The eyes have a narrow black mask outlined in white. The secondary wing tips often are a waxy red. Legs are black. It is 6-7 in. long with a 9-12 in. wingspan. Males and females look alike. They are nomadic over North America, breed in more northern areas, and wander widely. We have seen them in large flocks during the winter in Connecticut, feasting on berries from *Rosa multiflora*, privet (*Ligustrum* spp.), and cedar (*Juniperus virginiana*). They prefer high-energy content food. My painting reflects the late summer diet.

Catesby painted this bird, calling it "The Chatterer":

> "IT weighs an Ounce; and is rather less than a Sparrow. The Bill black; the Mouth and Throat large…What distinguishes this Bird from others, are eight small red Patches at the Extremities of eight of the smaller Wing-feathers, of the Colour and Consistence of red sealing Wax" (I-46).

Morrow's honeysuckle is a multi-stemmed, deciduous shrub that grows to 7 ft. tall. The leaves are opposite, 1-2 in. long, oblong, on short stalks, lightly hairy above, very hairy below. Arising in pairs from the leaf axils, the flowers are white and tubular with 5 petal lobes. The fruits are red to orange, many seeded, maturing in mid-summer and persistent through the winter. It is naturalized in the eastern half of the United States from Maine to North Carolina. It will occur wherever land is disturbed, its seeds dispersed by birds and mammals.

The cedar waxwing was painted from a specimen from Connecticut College, kindly lent by Doctor Robert Askins, Professor of Biology. Morrow's honeysuckle was painted from roadside specimens gathered in Connecticut.

in browsing the plant in Europe) is no longer present, resources may be shifted to another of the plant's functions, such as growth and reproduction. So, when tested with native European root-feeding weevils, a natural herbivore in Europe, the insects were larger and survived better on North American plants. It is likely that some defense against the insect was weakened in the North American plants and the resources previously lost to the insect (or put into the re-growth of roots) were put into reproduction. The thought is that when a plant invades a new area, pressure from native herbivores is reduced and selection favors a shift from defense to other processes that promote the species' survival.

On the other hand, shoot mass of North American plants tends to be greater than in European plants. North American purple loosestrife plants have more phenol-like compounds than their European ones. Phenols are good insect repellents; however, larval development and adult size of foliage-feeding insects did not differ in plants from the two continents. Further, large Saturniid moths, which are generalist herbivores, have begun to feed on purple loosestrife in some areas in North America.

Control of purple loosestrife is still problematical. In small infestations, young plants can be removed by hand before seed is set. Herbicides such as Roundup™, a glyphosate, are effective, but expensive for large areas. Biological control is the most efficient practice in the long term. Accordingly, a species of a root-mining weevil and two species of leaf-feeding beetles have been approved for use in purple loosestrife control efforts. Two other species of flower-feeding beetles are also being tested. Generally, the insects will feed exclusively on purple loosestrife and survive only as long as the plant. They will not attack native species, which, if all goes as planned, should return to their former range.

IV

WHERE THE LAND MEETS THE SEA

Salt Marsh: A Dynamic Equilibrium

Primary productivity in salt marshes is among the highest of all plant communities. Further, a salt marsh is built by the organisms that inhabit it, a process called autogenesis (see Oysters, pg. 119). Salt marshes are also unusual in that much of the biomass produced is not used directly by other organisms, but is converted to detritus. It is this plant material, and the associated microorganisms and invertebrates, which are the primary food source for both resident and transient consumers. The growth patterns of the dominant marsh plants provide habitat stability. In addition, the plants remove the chemical pollutants that cause eutrophication and convert these compounds to nutrients. Marsh plants are also capable of processing organic wastes, which provides water treatment over large drainage areas. Salt marshes are important stopover sites for migratory shorebirds and waterfowl, and supply food and protection for resident birds, mammals, fish, and a large array of invertebrates.

Wetlands offer spawning, nursery, and feeding areas for commercially important fish species, shellfish, and crustaceans. Marsh plants tolerate oxygen-depleted soil and high salinity; the above ground stems and leaves slow water movement and facilitate sediment deposition. The aggressive rhizomes lead to rapid colonial expansion. Seeds dropped in less brackish estuarine water germinate rapidly and are less likely to drift off on an ebbing tide into more saline conditions. Germination is delayed in seeds that are moved away, and the chances they may lodge in an unvegetated sandbar or mud flat are increased; so is the possibility that new salt-marsh areas will be established. Salt-marsh dynamics are tied into longer process as well. Periodic rises in sea level, for example, wash away old barrier islands and drown tidal lagoons. The eventual emergence of new salt-marsh with the continued riverine deposits of soil and minerals illustrates a millennial-long dynamic between the marsh and the sea. The ability of a salt-marsh to rise as fast as or faster than sea level lies in the stability of the substrate created by the plants landward root system. The root complex provides continuity, and sod deposit needs only outpace the twice-daily tidal intrusion to grow and expand.

Salt marshes can persist for thousands to tens of thousands of years. On the Atlantic coast, salt marshes are dominated by two cord-grass species: *Spartina alterniflora* (Plate 15, pg. 91) and *S. patens*. Although closely related, the two species differ in their ecology and distribution. *S. alterniflora* is the pioneering form and its seeds will germinate even after submersion in seawater for many weeks. The taller, *S. alterniflora* marsh cannot extend more than about a foot or two above mean high water because of the presence of its congener, *S. patens*. Rarely higher than 1-2 ft., *S. patens* (called salt hay in New England, salt meadow hay in the Mid-Atlantic States, and marsh hay cord grass in the South) creates a dense, impermeable mat of materials that, where it is most lush, excludes all other species. In Catesby's time, salt hay is harvested twice a year. Cut young it served as a high protein livestock food. Older hay was used to insulate walls and for roofing, and to 'bank the house' by insulating the foundation to keep out drafts. Because salt hay will not seed itself unless salt is present, older salt hay afforded the farmer mulch without weeds, and is still used for that purpose today.

The mixing of the silt-bearing streams and tidal wavelets provide ideal growth conditions for these two plant species. The spoils from dredging activities have become a prime source of real estate for even further expansion of *Spartina* marshes. The estuaries of the Chesapeake and the coast of Georgia are the sites of their most vigorous growth and development, but salt marshes are widely distributed especially south of Delaware to the Florida Keys. Of course, at the same time, marshes are dredged and filled to form new dry land for roads, parking, building and recreation, all of which increase susceptibility to storm damage and flooding, and reduce the marsh's natural functions.

In established areas the health of the marsh depends less on seeds than on rootstock. *S. alterniflora* dies back each winter and only coarse stubble remains. In spring, fresh grass emerges from the old rhizomes. The lushness of the growth is determined in great part by the abundance of rainwater or estuarine mixing. Winter dryness can lead to stunted growth. A suffocating oil spill or the herbicidal effects of environmental chemicals may inhibit growth and renewal. Fortunately, the buried rhizomes are resilient and unless an impervious crust forms, salt marshes usually recover in a year or two. It takes longer, years to decades, for the more subtle relationships among other plants and animals living in the *Spartina* matrix to redevelop.

While the two *Spartina* species are elemental to salt-marsh ecology, other plants occur and salt marshes are highly zoned. Discrete, single species bands dominate specific elevations and unvegetated spots are rare. Unusually high tides may deposit old *Spartina* cane on top (landward) of a marsh hay meadow which itself represents a detritus rich, elevated sod. This provides a rich growth medium and is protected from exposure to brackish water. Here, black needle rush (*Juncus roemerianus*) forms dense mats that resist both further *Spartina* invasion and erosion. Saltwort (*Batis maritima*), also called turtle weed, is another native plant highly adapted to a saline environment with seeds that can sprout after several months' exposure to seawater. Turtle weed can survive direct exposure to hurricanes and quickly recolonizes. The flowers, incidentally, are attractive to butterflies. Sea oxeye (*Borrichia frutescens*) inhabits the higher reaches of salt marshes from Virginia south. Although seeds are produced, it too propagates primarily by rhizomes, which results in dense colonies that compete with *S. patens* where permitted by even an additional few in. of elevation. As with the other plants of coastal marshes and dunes, Sea oxeye tolerates sediment buildup and affords resistance to erosion.

A major threat to salt marshes along the eastern seaboard is the invasive common reed, *Phragmites australis*, which has aggressively occupied thousands of acres of tidal wetlands. Although handsome as a decorative item when dried, it invades, spreads rapidly, and excludes native plants and animals. Habitat quality is rapidly compromised. Successful invasion has been encouraged by man-made restrictions to tidal flooding patterns, by the filling of marshes, and the degradation of water quality. *Phragmites* dries seasonally to become a potential fire hazard. Control is difficult in large part because of the resilience of the plant's rhizome system and the expense of restoration of tidal flow patterns or remediate filling.

Both upstream damming and the diversion of fresh water tip tidal salt marsh dynamics toward decline. Government actions to protect, restore and preserve salt marshes have met with varying degrees of success. Despite wide-ranging efforts at conservation, cord-grass salt marshes along at least 1,000 miles of the south-east and Gulf coasts have experienced a rapid, unprecedented die-off in recent years. It is highly likely that a severe drought may have weakened the plants either directly or through changes in the soil such as increased salinity. But the periwinkles (*Littoraria irrorata*) that graze on the plant are also implicated. Where they occur in high numbers their feeding mechanically harms the plant, which facilitates fungal invasion, in turn leading to plant death. The numbers of these snails which are a normal part of the food web increased due to reduced numbers of blue crabs (*Callinectes sapidus*, Plate 19, pg. 105), a major predator on *Littoraria*. The combination of disturbed trophic interactions, climate change, and human disturbance all influence the health of the salt-marsh.

The ecological role of *Spartina alterniflora* in marshes along the south-east and Gulf coasts is indisputable. Conditions shifted when *S. alterniflora* was accidentally introduced in the West where it freely interbred with the related California cord grass (*S. foliosa*). The resultant hybrid plants are taller and grow lower towards the low-water mark. Because the hybrid can self-pollinate it spreads more aggressively than the native west-coast species. Seeds of *S. alterniflora* were probably carried west on oysters shipped by train following the California gold rush, then to the North-west as the oyster business expanded. *S. alterniflora* can be controlled by weed killers, but a better option is biological control. In the East an aphid-like insect, a planthopper (*Prokelisia marginata*), provided natural control. However, in the new environment the plants that have not been exposed for generations vary in their ability to resist and tolerate the planthopper. While mortality in many plants may be high, in some populations resistant plants remain. Because control is incomplete, the population may eventually recover. Meantime, the unwelcome changes to the marsh and its inhabitants persist.

IBISES: WHITE AND SCARLET

A small flock of the white ibis (*Eudocimus albus*, Plate 16, pg. 95) settled into a pond on the west end of Shackleford Bank, just off Beaufort, North Carolina. They come regularly in March and all appear to be in adult plumage. White ibises are common in coastal salt marshes, swamps, mangroves, and the shores of lagoons and shallow lakes. In colonial times, in eastern North America, they may have been more abundant in grass and sedge wetlands. Ibises have habituated to humans and can be found foraging around outdoor eateries in south Florida and through the Keys. Normally they forage head down, walking slowly and probing the mud with the long curved bill. In sunlight the plumage is intensely white, with greenish-black wing tips, set off

Smooth Cordgrass
(*Spartina alterniflora*)

Ribbed Mussel
(*Geukensia demissa*)

Olive Nerite
(*Neritina reclivata*)

Saltmarsh Periwinkle
(*Littoraria irrorata*)

Smooth cordgrass, also known as salt-marsh cordgrass, is an herbaceous, native, warm season grass adapted to the salt marsh habitat. The short form grows to 16 in. and the tall form to 8 ft. with stiff stems surrounded by long narrow leaves. The inflorescence is spike-like. It reproduces by seed or underground rhizomes. Because of its value as an ecosystem builder it has been transplanted worldwide.

The ribbed mussel grows to 4 in. in length. The shell is oval, grooved, with a narrow blunt head. Shells are olive-brown to dark brown with yellow and white markings and iridescent white with purple edging on the interior. It attaches to the rhizomes of *Spartina* with a byssal thread. Mussels also filter water and provide nutrients to the *Spartina*. It is found from Maine to the Gulf of Mexico and was introduced into San Francisco Bay.

The olive nerite is a smooth globular snail to 1/2 in. in diameter. It climbs *Spartina* stalks, feeding on algae and macroflora. Nerites are popular aquarium snails.

The saltmarsh periwinkle grows to about 1 in. in length. It is a dull gray-white with dashes of red brown on the spiral ridges of the shell which is composed of eight to ten whorls. The aperture is oval. *Littoraria irrorata* occurs along the coast from New York to Texas. While saltmarsh periwinkles do not feed on *Spartina* directly, they can cause the plant considerable damage. They puncture the blades of grass with their radula, consuming fungi that then grows on the scar.

I painted these specimens from the marshes in Beaufort, North Carolina, near the Rachael Carson Reserve.

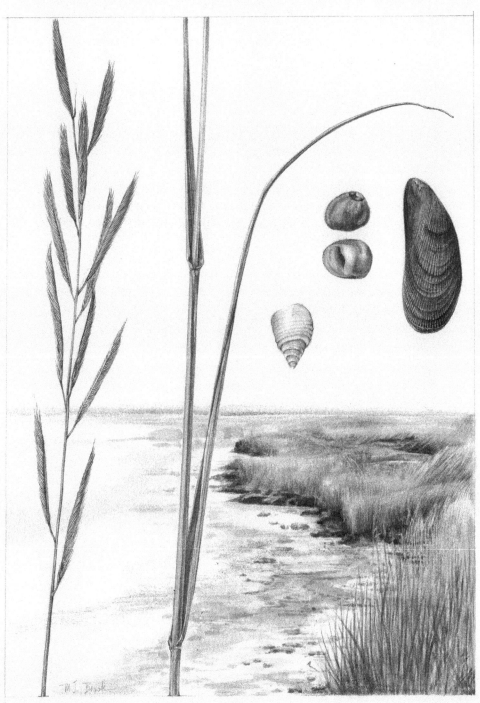

neatly by the red-orange bill and feet. Males and females differ in size but not in plumage.

Despite its lanky appearance, long bill and legs, a white ibis weighs about 2 lbs., no more than a gadwall (*Anas strepera*) and less than a mallard (*Anas platyrhynchas*). Catesby described the white ibis as "…about the Size of a tame Pigeon." This is hard to reconcile - the 25 in. long, 2 lb. ibis with the 12 1/2 in., 4.2 oz. feral pigeon (*Columba livia*). Catesby did note that the flesh and especially the fat "is, very yellow, of a Saffron colour" (I-82). Catesby illustrated the adult as "The White Curlew" on one plate, and the juvenile as "The Brown Curlew" on another, but noted the similarities in size and structure. The name curlew was most likely assigned because of the similarity in the long, down-curved bill to that in the Eurasian curlew (*Numenius arquata*) which Catesby must have known. He reports dissecting birds primarily to sex them as he first thought these plumage differences in adult and immature were sexual differences. He obviously collected more than a single bird and also recorded that the gizzard contained crawfish. In addition, the diet includes small fish and amphibians, insects, and mud-dwelling invertebrates. This diet is energy-rich and a wonderful source for carotenoids, the chemical that colors the skin of the face, foot, bill, and internal organs.

White ibises roost and nest in wooded swamps and flooded fields along the coast of North Carolina and southward along the Gulf Coast into Central America, the Greater Antilles and the coast of South America. They occur only rarely in the Bahamas and do not breed. Catesby mentioned that they "come annually about the middle of *September*, and frequent the watery *Savannas* in numerous Flights, continuing about Six Weeks, and then retire, and are no more seen until that time next Year." He claimed "In many of the Hens… (there) were Clusters of Eggs; from which I imagine they retire somewhere South to breed" (I-83). This implies they breed elsewhere, but the seasonal cycle is unclear. Catesby pointed out the autumn weather (late October) "…would probably be too cold for that Work of Nature, it being much colder in the same Latitude in that Part of the World than in *Europe*" (I-83).

Catesby's "The Red Curlew" is now known as the scarlet ibis (*Eudocimus ruber*, Plate 16, pg. 95). His figure of the scarlet ibis differs from most others in that no plant was included. Catesby mentioned that the scarlet ibis was larger than the white, "being about the bigness of a common Crow" (I-84). This, too, is certainly in error on two counts. The current evidence is that the two ibises are essentially identical in size, and both are larger than either the American crow (*Corvus brachyrhynchos*) or the carrion crow (*Corvus corone*) of Europe. The scarlet ibis (*Eudocimus ruber*) is a bird of the tropics, preferring coastal swamps, lagoons and mangroves. The distribution of the scarlet ibis makes it highly unlikely that Catesby collected his "Red Curlew" in the Carolinas. It is a resident of the northern coast of South America. It has been reported rarely (an accidental) only along the coast in Texas, Florida and Alabama, but overlaps with the white ibis in Venezuela. Nevertheless, the precise relationship between the two ibises is not settled. They may be alternative color phases, separate subspecies, or different species. The matter will eventually be settled by laboratory studies and adjudicated by a committee of the American Ornithologists' Union. In any case, a traditional test of the matter of separate species is hybridization. By one definition two populations are the same species if, when given the chance, they hybridize and produce viable offspring. The red ibis and white ibis hybridize in captivity and produce viable young. But hybridization under natural conditions occurs only very rarely and little is known of the hybrids. In areas where the two overlap there is no evidence as to whether the two forms mate randomly or prefer mates of the same color. Called assortative mating, the term implies they may be different species. While the jury is still out on the species issue, the plumage colors of the two raise some interesting questions and the biology of carotenoid pigments is an area of intense interest.

The pigments responsible for the yellow and red colors in many birds are carotenoids, a well-studied group of chemicals. Carotenoids are fat soluble, long-chain, organic molecules quite widely distributed in nature. They are synthesized only in plants, for example carotene in carrots. Animals obtain them from their diet but can modify them metabolically. Crustaceans (the reds in shrimp and lobsters for example), insects, other invertebrates, and even fruit and algae are good carotenoid sources for birds. Once ingested, carotenoids are transported in the blood, and dispersed in tissues, especially fatty tissues including the liver and fat bodies. In some, but not all birds, carotenoids are transferred to the skin and feathers where they produce brilliant colors. The two *Eudocimus* species are an extraordinary case in point. The chemical modifications are such that the intensity and hue of the colors are changed. Catesby's observation of yellow (saffron) flesh and fat bodies in the white ibis indicated they ingest significant amounts of carotenoids, but do not deposit the pigments in the feathers. The red color of the cover of the beak, legs and feet is produced by carotenoids, most likely derived enzymatically from the yellow precursors. By contrast, the scarlet ibis, with a very similar diet, does indeed deposit significant amounts of carotenoids in the feathers as well as the beak and legs.

The fact that one species deposits pigments in its feathers and another, very closely related one does not, even when the pigment is known to be present in the diet, is something of a puzzle. Further, both have reddish legs and intensely carotenoid-pigmented bills. It is known that carotenoids, or their precursors, are absorbed from the gut and then transported by the blood to the various tissues. The presence or absence of a carrier molecule in the blood might determine the fate of the pigments. The scarlet ibis has a serum protein, absent in the white ibis, which binds and transports carotenoids. Both species deposit pigment in various internal tissues. This means the selective step might be at the feather itself, but no precise mechanism has been reported as yet. In other species there are both sexual and seasonal differences in plumage colors. Chemical conversions of pigments are executed enzymatically and can be controlled hormonally. In some instances the plant pigments are deposited unmodified in feathers. The processes of incorporation, modification, transportation, and fractionation of carotenoids are yet to be fully elucidated.

White Ibis
(Eudocimus albus)

Scarlet Ibis
(Eudocimus ruber)

Black Mangrove
(Avicennia germinans)

The white ibis is a medium-sized wading bird 2 in. tall with a 3 in. wingspan. The plumage is entirely white except for the black tips on the wings, seen best in flight. The red-orange bill is long and strongly downcurved. The face is also red-orange, devoid of feathers. The legs are long and pink-orange. Males and females are similar in coloring, but the male is far larger in size. White ibis is distributed along the Atlantic and Gulf coasts from Virginia to Central America. Catesby called this bird a "White Curlew".

The scarlet ibis is the same size and wingspan as the white ibis. The body contour feathers, face and legs are brilliant red to pink-red depending on location. Wing tips are black. The bill is black and down curved. They occur on salt-water flats along the coasts of South America from Argentina and Brazil with vagrants in the Yucatan and the United States.

Black mangrove (*Avicennia germinans*) has long horizontal roots. It is a tropical tree that ranges from the equator to 28 degrees North and South. It grows in coastal areas, bays, lagoons and tidal creeks above the high tide line. In Florida from the Keys to St. Augustine, it can reach heights to 50 ft. Leaves are green, opposite, 2-4 in. in length, narrow and often encrusted with salt. The upper sides are shiny and the underside hirsute. The bark is dark brown and scaly. Small blossoms are creamy white and appear in June.

I observed these birds in St. Augustine, Florida and painted from study skins from The American Museum of Natural History, New York City. The mangrove specimen came from the Merritt Island National Wildlife Refuge in Florida.

Great Blue Heron

The great blue heron (*Ardea herodias*, Plate 17, pg. 99) is a large, lanky bird that typically frequents both coastal and fresh-water marshes. Its posture is described as dignified and its movements deliberate. The plumage carries a distinctly formal aspect. Perched or standing on the ground, the great blue heron demands respect, especially as it strides slowly on the water's edge to search for prey. In very cold weather, ankle deep in snow, it projects an attitude of trying not to look cold, but somehow does not quite succeed. These herons range widely in North America.

Catesby observed "The Largest crested Heron" both in Virginia and the Carolinas and probably also in the Bahamas. He estimated its size at "not less than four feet and an half high, when erect" (App-10).

The long legs and neck coupled with a 72 in. wing span make this a "great" bird indeed. In fact, these dimensions make it possibly the largest bird most casual observers will see. Commensurate with its size and bearings, it sounds a deep '*frawnk*' when flushed.

Catesby noted, especially, "The bill measured almost eight in. from the angle of the mouth to the end of it" (App-10). Not surprisingly, the beak is conspicuous in Catesby's figure. The avian bill, the jaws with their horny sheath, is a telling feature of any bird species. The beak is the major tool through which birds manipulate their environment. The wings of birds are given over completely to flight, or the occasional threat, or mating display. Beaks are the primary tools for preening, an activity critical to maintaining the plumage. The great blue heron has an additional tool for plumage maintenance. The pectin, a comb-like structure on the large claw of the middle toe, is certainly capable of making fine adjustments to feather structure and may also be helpful in removing feather lice. It is found in only a few other, unrelated species.

The beak is indispensable in feeding, which can involve grasping, holding, tearing, crushing, pecking and whatever mechanical action might be required as a replacement for our hands and tools. Birds, like primates, are erect, bipedal animals. The hind limbs provide ground locomotion and support while standing. Many birds can stand on only one foot. Raptors may use the feet for grasping, others use them for scraping, but these are exceptions. In other bipeds, raccoons, bears, monkeys for example, the hands serve multiple functions, even without a completely opposable thumb. Consequently, in birds the bill has assumed most of these roles, and more. This is reflected in the enormous variety of bill shapes ranging from the massive, hooked beak of eagles to the delicate, needle-like precision instrument of warblers. Bill shape reflects function and is intimately related to feeding, gathering nesting materials, exploring the environment, and even social interactions, including the dramatic mating displays in albatrosses. The eight-in. spear of the great blue heron is only one example of this wide array.

Like many other water birds and shorebirds, the great blue heron, on its very long, naked legs, often belly deep, wades in search of prey. They feed primarily on fish but also take frogs, invertebrates, lizards, and occasionally even small mammals, especially when ponds and marshes are frozen or snow covered. They hunt by walking slowly or standing seemingly immobilized, in wait of prey. There is evidence that great blue herons have excellent night vision, which would extend their foraging time. The great blue heron thrusts its neck rapidly forward, using its bill to grasp the prey whole, or to spear it, and then swallows it whole.

Based on evidence from numbers and distribution, the great blue heron has done well in recent times. Coincidently, populations of the closely related gray heron (*Ardea cinerea*) are doing well in England and Wales. At the turn of the nineteenth century, the great blue and other herons were seriously threatened by hunters who gathered eggs for food, or shot adults in breeding plumage for the feathers used in the millinery trade. In the early twentieth century the US Congress initiated protective measures through legislation. In concert with private organizations, these actions launched the conservation movement in the US, led by the National Audubon Society and the American Ornithologists' Union. The threat in this case was not from habitat loss, as has affected other species, but from intense commercial hunting pressures.

The future, however, is still not secure. In eastern North America the great blue heron is associated with wetlands from Nova Scotia and coastal Maine all the way to Florida. Colonies exist on islands in the St. Lawrence westward across the continent. Because breeding colonies need not be located in, or directly adjacent to feeding grounds, there is great latitude in where colonies are established. Nests generally are constructed in trees, but they also are built in shrubs, on the ground, and on cliff ledges. Some colonies, especially those in large stands of trees, are heard or smelled before they are seen. Along the Atlantic coast, great blue herons feed in riparian swamps and fresh water marshes, and it is these wetlands that are at risk. Efforts to preserve and restore such habitat are effective, but need to be expanded. Preserving wetland habitats will be critical to their survival.

Green Heron

The green heron (*Butorides virescens*, Plate 18, pg. 101) favors swamps, wooded wetlands, ponds, river edges, lagoons, and mangroves. It is secretive when feeding at night, but is easily observed by day in urban settings where it stalks at water's edge. It often perches on docks, mooring balls, jetties, and boats. It can be bold enough to appear regularly for handouts at fishing boats as boat crews clean the catch. It is a solitary nester and only rarely breeds in small groups. The flimsy nest is built of twigs and placed in low bushes or trees overhanging water.

Catesby called this species "*The Small Bittern*." The generic name *Butorides* means "resembling a bittern." Bitterns are a group of small marsh birds related to herons, similar in morphology, but much more secretive, albeit with much louder calls. Catesby's description of the green heron is accurate in regards to size and proportions, especially the beak and feet. The beak is relatively massive for the body size allowing it to capture a variety of prey. Catesby also noted that they were not seen in Virginia and Carolina in the winter and suggested "…they retire…more South" (I-80).

Green herons wander after breeding, often dispersing over great distances. This behavior can gradually merge into a migratory pattern. Still, the exact migratory routes are poorly known.

GREAT BLUE HERON
(*Ardea herodias*)

The great blue heron stands 4 ft. tall with a 6 ft. wingspan. The plumage is a grayish blue on the upper body with chestnut colored streaks on the thighs. A plume of black feathers starts behind the eyes and extends back to the crown of the head. The neck is long; the legs are long and yellowish-grey. It ranges from southern Canada to the West Indies and Mexico. It also occurs on the Galapagos Islands.

Catesby called this bird the "*The Largest crested* HERON". He wrote:

> "AS I did not measure the length of this Bird, I can only guess it to be not less than four feet and an half high, when erect…The crest on its head was made up of long narrow brown feathers, the longest being five inches in length, which it could erect and let fall at pleasure. The neck and breast brown, but paler, and spotted on the under-part. The rest of the body and legs brown, except the quill feathers, which are black. They feed not only on Fish and Frogs, but on Lizards, Ests, & c." (II-Appendix-10).

I have painted what I think to be the most striking features of the great blue heron, the head and long serpentine neck.

Green Heron
(*Butorides virescens*)

Green heron adults measure 18 in. in length with a 26 in. wing span. Other common names are green-backed heron, striated heron (now applied to a different species) and shitepoke in reference to its fecal projectiles. It has a greenish cap with a small crest and an iridescent gray-green back with chestnut brown below its eyes that extends to the upper chest and sides of neck. The tail is blue-green above and greyish-white below. The feet and short legs are orange-yellow. It is common from southern Canada to northern South America generally near wetlands.

Catesby called this bird "*The Small Bittern*" and described its coloring as:

> "The large Quill-Feathers of the Wing of a very dark Green, with a Tincture of Purple. All the Rest of the Wing-Feathers of a changeable shining Green, having some Feathers edged with yellow." [Catesby was describing iridescence.] "They have a long Neck, but usually sit with it contracted, on Trees hanging over Rivers, in a lonely Manner waiting for their Prey, which is Frogs, Crabs, and other small Fish" (I-80).

I painted this watercolor from a bird in Carolina City, a stop on the Intracoastal Waterway during our journey south.

Spring migration to breeding areas occurs generally in late winter or early spring depending on latitude. Green herons usually arrive before the larger heron species. It is thought that its twilight feeding habit may extend its day giving it a head start on breeding.

Green herons are reasonably tolerant of human activity. They were once shot for food and have suffered the effects of pollution, especially the wide use of insecticides. Nevertheless, no population is currently threatened despite this and changes in their natural habitat. While recreational use of river channels and swamps discourages green herons, they seem to adapt well to backwaters and marginal pools. The fate of many heron species now seems tied to conservation and restoration of these low-lying flooded areas. There may be longer-term threats from climate changes, in wetland drainage patterns, and drought. The damaging effects of sea-level change will not be a direct result of the gradual rise predicted from the melting of glaciers, but through extreme events such as floods and storms. The magnitude of the impact of such events will depend on the vulnerability of the local coastal zone. An intense hurricane season will do more damage, more quickly, than a rise in mean sea level.

It is already apparent that changes in climate have an effect on birds. The seasonally recurrent events in the lives of birds, termed the phenology, such as the return of a migratory species, date of first laying, last sightings, and molt pattern, are subject to meteorological factors and therefore climate change. The timing of migration and nesting seem to be particularly malleable. Many species in the temperate north are returning earlier and nesting earlier than in years past. This is associated not with changes in day length, but is tied to mean temperatures. The increase in temperature, of course, is the result of global changes. Day length and air temperature are called proximal factors in the cycle. Seasonal day length is a much more reliable clue than mean daily temperature. In fact, the response to photoperiod may be subject to natural selection, and has changed as climate changes. The consequences of alterations in phenologies on the dynamics of the populations are currently of intense interest.

On The Beach — Some Crabs

Catesby described coastal Carolina as:

> "…low, defended from the Sea by Sand-banks, which are generally two or three hundred Yards from low Water Mark, the Sand rising gradually from the Sea to the Foot of the Bank, ascending to the Height of fourteen or sixteen Foot. These Banks are cast up by the Sea, and serve as a Boundary to keep it within its Limits. But in Hurricanes, and when the strong Winds set on the Shore, they are then overflowed, raising innumerable Hills of loose Sand further within Land, in the Hollows of which, when the Water subsides, are frequently left infinite Variety of Shells, Fish, Bones, and other refuse of the Ocean" (I-ii).

This description still fits much of the coastline today and is precisely what we saw in morning walks along mudflats on Cumberland Island, Georgia, or around Beaufort, North Carolina. The effects of the sea remain just as potent today in shifting sands, depositing treasures, and destroying man-made structures as they did 300 years ago.

The beaches, tidal flats, and salt marshes along the Carolina coast support large crab populations. For as long as there have been tidal mudflats, fiddler crabs have occupied the seashore. On hot, sunny days, hundreds can be seen scurrying over the mud, pausing to pick a microscopic mote off the surface or diving into a convenient hole. Three species of the genus *Uca* occur along the Atlantic coast and in places their habitats overlap. The mud fiddler (*U. pugnax*) is most common in brackish and saltwater marshes. The Atlantic sand fiddler (*U. pugilator* Plate 19, pg. 105) shows a preference for sandier conditions and is generally found higher on the beach. The brackish-water fiddler (*U. minax*) tends towards freshwater to mildly brackish marshes. When they are not searching for food, fiddlers work continually to maintain their burrows, which provide a protected site from which to scavenge during the day, shelter at night, escape from predators and a retreat for the winter.

Fiddlers form large colonies and during the day a particular patch will seem to be literally crawling with crabs. However, in contrast to many other animal colonizers such as bivalves, a small inter-individual distance is rigorously maintained. Fiddlers feed on detritus that includes decomposing plant material, minute algae, and bacteria. Occasionally they take tiny crustaceans and are at times, though infrequently, cannibalistic. They feed on most surfaces, in sediment, in the wrack line and flotsam, and select patches of beach rich in resources. As food is depleted they move on to other, more profitable, patches. This behavior explains, at least in part, their seemingly frantic activity.

The massive claw of the males serves in both courtship and territory defense. While fighting activities are mostly ritualized and little damage is inflicted, the cost in energy is high. The enlarged claw, which can account for 40% of the body mass, is useless in feeding. Males feed only with the small claw. While this is not exactly trading manliness for food, it does entail significant constraints. Males feed only half as efficiently as the small-fisted females, because feeding is limited by the ability to scoop food into the mouth. Females are capable of ingesting food at twice the rate of males, and are limited mainly by the rates of extraction of food in the gut. Consequently, males, with their reduced capabilities, tend to feed in the richer sediment patches. To maximize food intake, males forage further from their burrow or vegetation cover. This involves greater risk of predation, but yields potentially greater rewards. The enlarged claw, useless for feeding or digging, finds its finest moment in the aggressive displays towards competitors and in courtship. Copulation in fiddlers as in many other crustaceans is preceded by elaborate courtship behavior. The role of the claw in reproduction is considered to be a product of sexual selection. The splendid tail of the peacock, the male peafowl (*Pavo indicus*) is another example of the same classic process of sexual display, which Darwin described as so important in signaling mate quality and preferred mate choice.

Crab mating rituals can be elaborate, especially considering the tiny size of the crustacean brain and nervous system. Typically female crabs can be impregnated only after molting. The reason is that the carapace and genital pore harden as the new shell hardens, preventing successful mating. The newly molted crabs of another species become the soft-shell variety of restaurant fame, consumable shell and all. As a consequence of the delicate nature of this timing, males may attach themselves to females days or weeks prior to their molting. In other words, males

WHARF CRAB
(*Sesarma cinereum*)

ATLANTIC MUD CRAB
(*Panopeus occidentalis*)

ATLANTIC SAND FIDDLER CRAB
(*Uca pugilator*)

BLUE CRAB
(*Callinectes sapidus*)

ATLANTIC GHOST CRAB
(*Ocypode quadrata*)

Wharf crabs (*Sesarma cinereum*), also called square-backed marsh crab or gray marsh crab, are semi-terrestrial and intertidal in habitat. Wharf crabs range from Maryland to the Caribbean, and are common on dock pilings, beaches, and in mangrove forests. The carapace is brown to olive in color. Males grow claws significantly larger than females. This specimen came from Swan's Island, North Carolina.

There are several species of Atlantic mud crabs, this being *Panopeus occidentalis*. They commonly grow to 3/4 in. in width, and have black tips on the large claws. Coloring ranges from beige to tan with varied markings in brown. This specimen came from mud flats on Jekyll Island, Georgia.

The tiny Atlantic sand fiddler crab (*Uca pugilator*) grows only to 1 in. in width. Males have one large and one small claw but in females, claws are equal sized. This watercolor shows a male top right and a female, just below, with eggs. Sand fiddler crabs range from Massachusetts to the Bahamas west to Louisiana. Common names are fever crab and calico fiddler crab.

Blue crabs (*Callinectes sapidus*) can grow to 9 in. wide and possess large blue claws that, in the females have a red tip. The carapace is a mottled brown. They are found along the Atlantic coast from Nova Scotia to Uruguay. I painted this crab in Carolina City in the cockpit of our boat, with a pelican overlooking the process. Needless to say, the aroma became stronger as the hot day progressed. Eventually, the pelican flew away with the just-painted specimen.

Atlantic ghost crabs (*Ocypode quadrata*) reach only 2 in. across and are light gold in color. They inhabit burrows along sandy beaches from Rhode Island to Brazil. Their conspicuous eyes, held aloft on stalks, make them easily identifiable.

I painted the Atlantic mud crab and the Atlantic ghost crab on Cumberland Island off the coast of Georgia while visiting Carol Ruckdeschel. Her research on sea turtles has been going on for many years, and she was an excellent source of information.

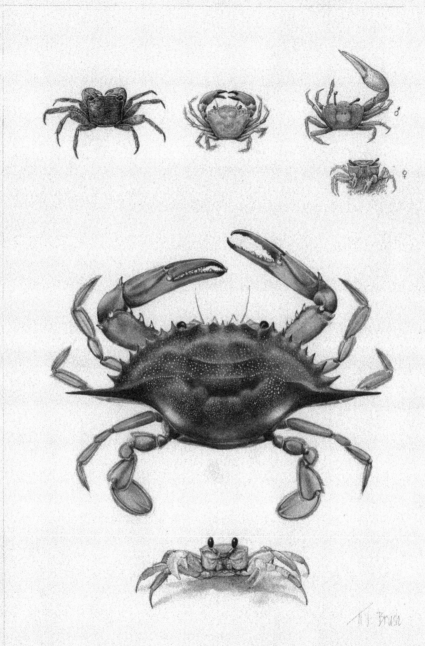

cannot predict, with accuracy, the internal condition of females that lead to the molt. During this period a successful male will use his claw to entice the female into the burrow where mating occurs.

After copulation, female crabs, like lobsters, carry the eggs until they are released as larvae. The free-swimming larva is subject to all the vicissitudes of life as a planktonic organism. Larvae are generally released at flood tide, which insures that they move rapidly away from the shallow water and mudflats, and the high risk of predation there. Larval behavior and morphology are often explained in terms of predator avoidance. Predators include plankton-eating fish and surface-feeding birds such as gulls and storm petrels. Moving off-shore reduces the larvae's exposure to some predators or the need for antipredator defenses such as elaborate spines. However, these waters are slightly cooler and typically nutrient-poor; hence they experience slower growth rates and endure longer development times than they might closer to shore. As with so much else in the lives of animals, the patterns represent trade-offs among imperfect alternatives.

FISH CROW

Crows, jays and ravens are ubiquitous, conspicuous, abundant, noisy and hard to ignore. Fish crows (*Corvus ossifragus*, Plate 20, pg. 109), are often the first birds heard on mornings in maritime and coastal areas. It takes only a few birds leaving their roost to call attention to the sunrise. This daily ritual has been repeated from time immemorial along the coasts and rivers of the South-east, and now more frequently northward, as fish crows began to nest in Connecticut during the early 1980s. Of all the North American corvids, fish crows seem most closely associated with water. Early writers mentioned that, like osprey (*Pandion haliaetus*) and the bald eagle (*Haliaectus leucocephalus*), fish crows will glide over water and use their claws to pluck prey from the surface. However, this feeding method has never been verified and remains in the realm of the mythological. As with other corvids, such as many ravens and jays, fish crows seem infinitely adaptable and are quick learners. The specific epithet, *ossifragus*, refers to their ability to crush bone or break open shellfish and indicates a favored means of feeding. Fish crows are also known to raid the nests of colonial water-birds, gulls, and terns. In urban areas they can be found in parks and other areas of human activity, generally feeding on garbage. Fiddler crabs on mud flats also are fair game, and the fish crow will occasionally eat carrion. In winter they add fruit and berries to their diet. Crows are quintessential omnivores. Both diet and behavior are molded to local conditions.

For whatever reason, neither the fish crow, nor the closely related and more widely distributed American crow (*Corvus brachyrhynchos*), were mentioned by Catesby. Lawson recorded the latter simply as crow. This name held for over a century. The earliest scientific description of the fish crow, published in 1812, is credited to Alexander Wilson. Wilson, born in 1766 in Paisley, Scotland, and trained as a weaver, was smitten with American birds the moment he stepped ashore in New Castle, Delaware. He eventually wrote, illustrated, and published *Ornithology*, the first work of its type in America. The illustration by Audubon, a contemporary of Wilson, appears to be the first illustration of the fish crow. It is chancy to speculate about why a species was not recorded by Catesby (or anyone else for that matter). It could be that the American crow was uncommon in the areas he visited, but it appeared on Lawson's list only a few years earlier. Alternatively, Catesby simply may have figured that the crow was so similar to the carrion, or Eurasian crow (*C. corone*) of Europe that it hardly needed mention. Thomas Nuttall, in his *Manual of American Ornithology* (1832), presumed the crow to be distributed almost worldwide. This is incorrect, but he did distinguish the fish crow, giving credit to Wilson for the original description. Nuttall mentions a strong association with water and their persistence around fishermen.

Winter roosts of American crows can comprise hundreds and even thousands of individuals. Fish crows do not roost in these numbers, but are no less conspicuous in their behavior. The two species are difficult to tell apart in the field. The physical differences reside in the relative size. *C. ossifragus* is slightly smaller overall, with relatively longer legs, a flatter forehead, and a slightly thinner, moderately hooked, bill. They are known to feed in mixed flocks that include both species. The two crows are most reliably distinguished by voice. The call of the fish crow is unlike the familiar *caaw* or *carrr* of the American crow. Listen for a shorter, much more nasal *carcă* or the two-syllable *că-hă*. They also offer a higher pitched, hoarser cahrr at times. The sound has been likened to an emasculated version of the call of *C. brachyrhynchos* and is instantly recognizable. Field identification is aided further if the bird is calling from an overhead wire, a posture rarely assumed by the American crow. In the case of the fish crow, it's all in the call.

The distribution of *C. ossifragus* in coastal and tidewater regions has expanded recently and its numbers have increased. Much of the credit for this expansion is related to the efforts to conserve and restore coastal wetlands. The adaptability and scavenging habit of fish crows has led to success in urban areas and other human-modified environments. As the food source has expanded so has the crow's range. This close association with humans has led to the implication of the fish crow, along with other birds, in the recent outbreak and spread of West Nile virus (WNV). The mobility of birds, their numbers, and close association with humans and mosquitoes, has produced a very rapid spread of West Nile virus.

West Nile virus first appeared in the eastern United States in 1999 in horses. Subsequently, it has been detected in about 140 bird species. In many areas the virus is common in crows and jays. Mosquitoes transmit the virus and once infected, susceptible species carry the virus in the blood for one to four days. West Nile virus has crossed the entire United States in less than three years and may be transported to the Neotropics by avian migrants. Many birds die from the virus, but others recover, and become immune. Birds are now the major reservoir for West Nile virus but incidental infections occur in mammals, especially humans and horses. Human symptoms are infrequently serious; four out of five infected individuals show no symptoms at all. Serious illness occurs in only one of 150 infected individuals. Recent reports now indicate that decreases in numbers of widely distributed, relatively common bird species such as the blue jay, black-capped chickadee, and American robin may be related to West Nile infection. Because mosquitoes spread the virus, there is the additional fear that it may be carried to the Hawaiian Islands. This is potentially devastating for the Islands' isolated and likely susceptible fauna. Avian malaria, also carried by introduced mosquitoes, has devastated many native species including the threatened Hawaiian crow (*Corvus hawaiiensis*).

Fish Crow
(*Corvus ossifragus*)

Fish crows reach 15 in. in length with a wing spread of 36 in.. Their plumage is glossy black and they have a thick black beak and relatively short black legs. Their distribution is along the coastal United States from Maine to Louisiana and inland up large river systems.

Catesby did not paint any crows, but he mentioned them while writing of the value of "Purple Martins" in the landscape:

> "The whole Bird is of a dark shining Purple; the Wings and Tail being more dusky and inclining to Brown. They breed like Pigeons in Lockers prepared for them against Houses, and in Gourds hung on Poles for them to build in, they being of great Use about Houses and Yards, for pursuing and chasing away Crows, Hawks, and other Vermin from the Poultry" (I-51).

I obtained this specimen from Dr. Robert Askins at Connecticut College, New London, Connecticut. As I have no chickens, I do not think them vermin.

AMERICAN OYSTERCATCHER

In 1605 Samuel de Champlain (1574-1635), during his exploratory trip along the east coast of North America (then called New France), found American oystercatchers (*Haemotopus palliatus*) in large numbers in what is now eastern Massachusetts. Historically, it is possible that they may have nested as far north as Labrador, but there are no records to verify this. There are records of oystercatchers being collected in Boston as late as 1837 and sold as food in local markets. By the mid-1800s though, the species had essentially disappeared as a breeding bird from the Northeast, leaving abundant numbers only to the south in Virginia and the Carolinas. The general consensus is that the fall in numbers resulted from over-hunting and egg-collecting. The oystercatcher's decline was not stemmed until the initiation of the conservation movement of the late nineteenth century and the passage of the Migratory Bird Act in 1919. Once protected, oystercatchers again expanded northward. Nesting resumed in New Jersey in 1947, in New York in 1957, and near Boston Harbor by 1968. The first nest in Connecticut was reported in 1981, and four years earlier on nearby Fisher's Island (New York). The American oystercatcher is now relatively common as a nester in Long Island Sound and along the coasts of southern New England.

American oystercatchers (Plate 21, pg. 113) are large, pied shorebirds with a bright red to orange bill, pink legs and a yellow eye-ring. Striking in appearance and noisy in flight, they are found along the Atlantic coast from Boston to the Indian River, Florida, and on Florida's Gulf Coast. They are restricted to maritime habitats as they specialize on marine mollusks for food and use coastal salt marsh and sand beaches for nesting. Despite habitat loss, over-hunting, and human encroachment, American oystercatchers have clearly benefited from the establishment of large coastal reserves along the mid-Atlantic seaboard. Further, the species has adapted well to dredge-spoil islands (another by-product of human activity) where they breed successfully. These habitat changes were only a part of the twentieth-century range expansion back into the Northeast. The expansion was preceded by increases in numbers of these long-lived birds (16-17 years) in Virginia and North Carolina. High dispersal capacity and low breeding-site fidelity facilitated the expansion.

The American oystercatcher is sedentary over much of its range. However, birds nesting in the Northeast move south as far as Central and South America during the winter. Individuals also wander in the non-breeding season, which may account for the breeding population in the Galapagos Islands. Even on these isolated islands, the oystercatcher inhabits habitat similar to their more familiar homes: rocky coasts and sandy beaches.

In long-lived birds such as oystercatchers, conditions of early development play a role in long-term fitness. Nests are simple scrapes in the ground, usually on a sandy surface. An average clutch contains three eggs, but clutches of up to six are sometimes recorded. Both sexes incubate the eggs and if a nest is destroyed they re-nest readily. Birds reared in high-quality habitat tend to settle in high-quality territories. The recent expansion of breeding birds into the Northeast almost certainly reflects a combination of conservation measures such as the availability of higher quality habitat and release from hunting pressure.

Catesby described the appearance and feeding behavior of the oystercatcher accurately and was the first to use "oystercatcher" as the vernacular name.

> "In Rivers, and Creeks near the Sea there are great quantities of Oyster-banks, which at low Water are left bare: On these Banks of Oysters do these Birds principally, if not altogether, subsist; Nature having not only formed their Bills suitable to work, but armed the Feet and Legs for a defence against the sharp edges of the Oysters….In the maw of one was found nothing but undigested Oysters" (I-85).

Beyond his description of this species Catesby essentially credits Pierre Bellon (1517-1564) as the authority for the use of *Haematopus* ("blood foot," all oystercatchers having red legs) as the generic name. Bellon was an early traveler in the Middle East. His writings included some of the first descriptions of the Egyptian pyramids as well as birds and mammals (including the imaginary unicorn).

Find a Spot; Settle Down

Every sailor fights a continuous battle against settling organisms. Marine waters are a rich soup of microscopic organisms in various stages of development. In warm waters along the Atlantic Coast settling begins as a sticky, fine film of algae and bacteria not easily wiped away. It happens quickly, practically immediately at the molecular level. A dinghy left at the dock for only a few days soon sports a layer of green, slimy growth. The problem is magnified on larger vessels with a much larger surface area below the waterline. Boaters call this fouling. Fouling occurs first along the waterline and the first few inches below because of the availability of light and slightly higher water temperatures. The prop, shaft and other unprotected metal parts also suffer. After a short time an unprotected hull will harbor seaweeds, sponges, bryozoans, and larvae of mussels and barnacles. In addition, one need only walk along a dock to see sea squirts, hydroids, bryozoans, amphipods, sea-stars, mussels, shrimps and crabs on the surfaces just below the water line. Sponges and barnacles are small organisms with only a simple nervous system. Both are highly adapted for an appropriate habitat. Many species depend simply on moving water for a food supply and a mechanism to hold them in place. Nevertheless, the processes of settling and growth that determine the nature of the fouling community, also present opportunities for their control. This is a rich example of biodiversity. Another familiar example is the thoroughly encrusted anchor every coastal seafood eatery has in the parking lot or by the door, matched only by the retired navigational aids used for advertisement. In fact, the surface of just about any object submerged in the ocean is subject to fouling.

Tiny drifting animals, plants, algae, and bacteria, especially cyanobacteria, unicellular diatoms, and the larvae of myriad species populate the microscopic plankton in the ocean's water column whose composition reflects the water's recent history. Most move passively with the currents. The fate of most plankton is to become lunch for something else, but there is safety in numbers. The density and mix of organisms present in a particular sample of water depend on various physical parameters. For one, local patterns of tides and currents determine water movement at

AMERICAN OYSTERCATCHER
(Haemotopus palliatus)

The American oystercatcher is a large, conspicuous shorebird, 17-21 in. long with a 32 in. wingspan. It has a long, straight, red bill, black back and head, white undersides and wingbar, and yellow eyes surrounded by a striking red eye ring. The webbed feet are light pink. Two species are recognized in North America. *H. palliatus* breeds along the east coast of North and South America, on Caribbean islands, and on the west coast from southern California to Mexico. It is also widespread in the Galapagos.

Catesby seemed to have a great respect for the structure of this bird:

> "Their Feet are remarkably arm'd with a very rough scaly skin. In Rivers, and Creeks near the Sea there are great quantities of Oyster-banks, which at low water are left bare: On these Banks of Oysters do these Birds principally, if not altogether, subsist; Nature having not only formed their Bills suitable to the Work, but armed the Feet and Legs for a defense against the sharp edges of the Oysters" (I-85).

The watercolor I have painted was based on sketches made while visiting the Galapagos. These birds are darker than our east-coast birds, but the behavior was the same.

any particular location. Water temperature will depend on the season and time of day. Salinity may be affected by rainfall and local runoff. Another factor will be the intensity of recent biological production, the organisms produced in the immediate areas. Buoyancy and viability affect dispersal but plankton move as passive particles with little control over their speed or direction. Many of these organisms eventually settle, and almost any surface may do.

The process of dispersal and settling may appear programmed because it seems so predictable, but colonization and settling are the results of random events. A soft, short-lived flora will likely arrive first. Small, simple, and fast-growing, it can change the surface texture and provides food for microscopic animals. The species in attendance may vary, but the groups involved are functionally very similar. The settlement of diverse algae, plants and animals is the natural state of things on rocks or reefs. While settling is common on boats, piers and other man-made objects, it also appears on the flukes of whales and on the carapaces of sea turtles. The process of dispersal is, in essence, a sweepstakes.

Dispersal insures that animals settle far enough from their origin so as not to compete with other individuals of the same species and to extend the distribution of the population. Most planktonic organisms are under a fraction of an in. in their greatest dimension, regardless of the size of the adult. This is just small enough that the laws of fluid mechanics determine position and progress in flowing water. Because the adult organism is sedentary or only slightly mobile, the microscopic larva is responsible for an action that will affect the entire life of the adult. Proper siting must provide the adult with a continuous and adequate food supply, potential mates, protection from predators, and an environment for constructing protective living quarters.

The earliest settlers may be followed by other seaweeds and almost certainly by herbivores and detritivores. The zooplankton includes amphipods and larvae of barnacles and mussels, which can remain in the water column for up to eight weeks before they settle and begin to graze. Similarly, sponges, tunicates, and hydroids appear. Unlike the planktonic organisms, these polyps and medusae have tentacles that give them some control over their locomotion, an edge in the race for a place. Given adequate light and water movement, a community of soft-bodied animals emerges, sometimes rather rapidly. Once attached, they filter food from the water and compete intensely for space. Growth and development continues as a more permanent attachment to the substrate is formed. Pioneers may favor a microhabitat to avoid the immediate surge of the waves, to reduce dehydration, and encourage behaviors that influences the pattern of subsequent settlement. Ultimately, a community develops.

Settling organisms are like weeds in the garden. They are perfectly healthy organisms, specialized to disperse and settle. Fouling is a human concept. Just as gardens present ideal habitats for weeds, it is equally ordinary for opportunistic, prolific marine organisms to prefer ready-made submerged surfaces. Marine organisms all over the world lead their lives attached to surfaces. They did so long before humans evolved, much less sunk pilings in the water. Humans have not been especially successful in outwitting either garden weeds or fouling organisms. Most of what happens in fouling is identical to natural settling in non-manmade portions of the environment. Many of the processes occur at the molecular level and are a function of the nature of the available surface and the presence of competing organisms. Communication is essential and often involves chemical information. Even small, relatively simple animals can both send and receive messages and respond in an appropriate fashion, that is, to settle or not. Boaters, the US Navy and marina owners search for solutions to these man-made problems. The most desirable solutions are reliable, eco-friendly, and cost effective. The animals, on the other hand, have been strongly selected to successfully settle, grow, and reproduce.

A similar process where organisms adhere to a hard smooth surface and eventually settle and grow describes precisely the formation of dental plaque. Metabolic by-products of the clinging organisms influence both the immediate surface and provide signals, which encourage other organisms.

Settling on a firm substrate, a rock or piling, is a highly successful strategy. Settling organisms swap a vulnerable, mobile juvenile life-form for reliable access to food and oxygen, effortless waste removal, and a mode of reproduction that involves large numbers of gametes broadcast into the environment. This strategy usually involves a holdfast which anchors the adult organism. The holdfast must resist the force of the current flow and removal by predators. Intertidal organisms must also withstand cycles of drying. Any adhesive mechanism must be capable of working underwater and, once fixed, it would be advantageous that no further energy be expanded for the animal or plant to retain its position. In some species the adhesion mechanism may facilitate local movement, as in the foot of a snail, the suction cup of octopus or the feet of seastars. Many species of mollusks, including the economically important mussels, adhere directly to the substrate or each other. Individuals, often in the tidal or sub-tidal zone, have access to constantly moving water, which brings both oxygen and nutrients. The downside is exposure to the varying physical forces of the moving water, desiccation when exposed to air and predation by crabs, snails, birds, and man.

The tidal zones of sandy beaches lack hard surfaces for settling sites. Yet many animals live here and they are subject to similar effects of tide, exposure, and predation. Many of the microscopic animals found in the tiny, but extensive and moist interstitial spaces among sand granules are filter feeders. Others exploit microscopic detritus deposits. Larger animals like sand dollars, a relative of sea urchins, also inhabit this stressful environment. They have the ability to burrow into the wet sand where their streamlined shape acts to keep the animal in place. The configuration of the shell is broadly analogous to the slotted wing of an aircraft where a pressure difference between the top and bottom surfaces generates a downward force that facilitates stability.

The presence of any living organism will change the substrate conditions and hence influence subsequent settlement. Commonly, the first organisms present in the marine environment are bacteria or algae. The ubiquitous red algae produce a protein, phycoerythrin, which attracts additional species. Signals of this type are not very species specific and may simply signify surface characteristics that are favorable for settlement. It is known that for many species to settle there is an obligate requirement for a biochemical signal. The signals may be conjugated proteins (that is, a protein combined with a polysaccharide), amino acids, or an amino acid derivative such as GABA. GABA is a chemical signal that acts in brains as a neurotransmitter. In addition to recruitment and settlement, other molecular messages may influence growth rates and developmental patterns. Direct contact by the larva to the surface is required, which implies that the chemical inducers are not released into the water where they would immediately be diluted or removed by currents. Settling is induced by random contact of the larva with the local algae or bacteria. The process is not dependent on light or vision but is chemosensory and analogous to human taste. Biologists realize that the recognition of a non-diffusing, surface-associated, contact-dependent, specific biochemical trigger that initiates a series of subsequent events is a "lock-and-key" mechanism. Such specificity is important as it controls the entire subsequent life-history and survival.

The existence of chemical messages in the external membrane of red algae implies receptors on the larvae. Indeed, the surface cells of various parts of invertebrate larvae are rich in receptor molecules matched specifically to the recruitment molecules. Further, there is another class of molecules, called transducers or "second messengers," which translate the message in the surface receptor into the interior of the cell to stimulate or regulate various processes within the cell. Two well-known such molecules are ATP and cyclic AMP, organic molecules that can store and transfer chemical energy. Both molecules and the associated metabolic process occur in all living cells and are considered universal in biology. There are, as might be expected, molecules that inhibit these processes. One of them, copper, blocks GABA-initiated settlement, and interferes with attachment and growth. It is no wonder that commonly used bottom-paints for boats are extremely high in copper. However, copper, in concentrations only slightly higher than those that prevent settling, is also toxic to swimming larvae not yet induced to settle.

An even better antifoulant is tin, which is no longer legal to use as it is more toxic than copper. There are other categories of molecules such as halogenated hydrocarbons that are much more specific in their antifoulant effects. While not immediately toxic, their long-term ecological consequences are untested and next to nothing is known about their persistence and accumulation in food chains. They are currently considered impractical for use as marine antifoulants. Other groups of molecules that show promise as antifoulants are structural analogues of GABA. These chemical relatives are designed to compete for the receptor sites, bind more tightly, and consequently prevent transduction. Various biologically derived glycoproteins can also selectively block membrane-based processes. One glycoprotein–lectin–has been suggested as an antifoulant, as have other molecules that block larval receptors or inhibit specific steps in recruitment and settlement. However, adequate technology for their use has not been developed.

There are other natural products that are potential antifoulants. Common fouling organisms share traits such as short generation times, relatively brief times as larva, and rapid growth and maturity. These traits are highly susceptible to interference and control. Because universal biological processes are involved (for example, membrane structure, and biochemical pathways that include receptors, signal molecules, growth factors and so on), methods of control may be quite general. Marine organisms themselves commonly produce protective compounds with anti-barnacle settlement properties. Other natural, non-toxic products such as silicone detergents may prevent settling because detergents strongly modify the nature of surfaces. Incidentally, detergents are routinely used as insecticides and bactericides and are available in your local supermarket. But problems with these molecules persist. Detergents, for example, have no long-term effect on the surface. Ancillary effects on other organisms are unknown, as is their potential fate in the environment. The insecticide DDT caused major environmental damage before its environmental and health consequences were full realized. The long-term health and environmental consequences of bioactive additives used in agriculture, foods, and medicines are poorly known and can have unanticipated consequences in humans through direct consumption and indirectly through their occurrence in drinking water.

Oysters

Along with the famous martini (shaken but not stirred), James Bond (aka 007) often requested a dozen oysters. Oysters may be the world's most famous bivalves. Oysters waltzed seaside for the Walrus and the Carpenter. Restaurants often feature oyster bars where experienced, raw bar attendants deftly shuck and serve the occupant raw, on the half shell, perhaps with a twist of lemon. Oysters Rockefeller are lightly broiled and grace tables worldwide. Oysters are popular and one unspoken implication is that consumption of oysters will somehow enhance sexual performance. Perhaps the image of oysters as an aphrodisiac is overrated and like many other things in life exists in the eye of the beholder. Bond (aka 007) would be astonished to learn that oysters are gonochronic alternate hermaphrodites that spawn and engage in external fertilization. Hardly the bad boy image of the world famous spy! Further, the smallest, most recently settled individuals function as males in their first year and undergo subsequent sexual reversal as they increase in size. Indeed, sex ratios in populations change over time and there is evidence that the sex and proximity of nearby individuals may influence the sex of any single individual.

In Catesby's words:

> "Amongst other Shells Sticking to the Rocks, environing these silent Waters, were Oysters, which stuck horizontally to the Sides of the Rocks, that Edge next the Hinge of the Oyster, being the Part fixed to the Rock. These following Kinds of small Shells sticking to Rocks, are never found in deep Water, but abide where they are covered, and uncovered at every Flux and Reflux of the Tide" (I-xliii).

The shell of the Eastern oyster (*Crassostrea virginica*) is extraordinarily variable and not especially visually pleasing, but well-armed with sharp protrusions (Plate 22, pg. 119). The colors are drab, the surface scaly and encrusted, and the valves asymmetrical. Inside the shell is smooth and luminous with shades of iridescent colors. Oysters also have the ability to coat irritating ingested particles with the same deposits, forming pearls. Nevertheless, aesthetics aside, oysters are easily recognizable throughout their 5,000 mile range. The shell is a masterwork. Each shell contains almost 70% (by weight) crystalline calcium carbonate ($Ca(CO_2)_2$). The organic material that provides internal scaffolding and the anchor for the ligament consists of up to 33% protein and small amounts of carbohydrate. The geometry of the calcium deposits reflects requirements of shape and stress. At a Rotary Oyster Festival in Swansville, North Carolina, we discovered that the oysters were trucked in from Texas as local supplies had been seriously diminished. We also learned that the local Carolina oysters had spiky shells as a result of lower calcium levels in the more acidic water. This made them more difficult to handle. The shells of animals from the Gulf Coast were smoother. We also learned how to shuck properly.

ATLANTIC BAY SCALLOP
(*Argopecten irradians concentricus*)
EASTERN OYSTER
(*Crassostrea virginica*)
WHITE SHRIMP
(*Penaeus setiferus*)
ROYAL RED SHRIMP
(*Pleoticus robustus*)
STONE CRAB
(*Menippe mercenaria*)

The fan-shaped shell of the Atlantic bay scallop (*Argopecten irradians concentricus*) grows to 3 1/2 in. wide and possess more than 14 radial ribs. Shells are symmetrical and have a mottled coloring in pastel shades of gray, brown, orange, red and yellow. They are found in seagrass beds from Connecticut to the Gulf of Mexico.

The two rough shells that enclose the Eastern oyster (*Crassostrea virginica*) are white to a gray tan in color. The extremely rugged shell is lined with an iridescent pearly coating that, when irritated with a grain of sand, will surround it with the same material to create pearls. They range from the Gulf of St Lawrence in Canada to the Gulf of Mexico. Catesby illustrated the "*Oyster Catcher*" bird (1-85), but not an oyster.

The female white shrimp (*Penaeus setiferus*) can also grow to 7 1/2 in. in length, the male slightly smaller. They are also called northern white shrimp, Daytona shrimp, lake shrimp and green tailed shrimp. The body is a bluish white, speckled with pink and black. The tail is green along the margins. Antennae are much longer than the body. White shrimp occur from New York to Campeche, Mexico. They are very popular with commercial fisheries.

The female royal red shrimp (*Pleoticus robustus*) can reach 10 in. long, the males are slightly smaller. Color can range from red to a milky white. They are found on the continental slope from Cape Cod to the Gulf of Mexico. Trying to find a good specimen for this watercolor was perplexing until I called Mrs. Jo Bomster of Stonington Seafood Harvesters in Stonington, Connecticut. She had the crew on one of their trawlers flash freeze one for me. The colors were natural and the specimen in excellent condition.

Stone crabs (*Menippe mercenaria*) can reach a width of 5 3/4 in. The upper carapace is speckled, brownish-red, legs a reddish-tan with a yellow band. The underside, or plastern, is cream colored. The larger crusher claw is reddish-cream with a black tip, the smaller pincer claw also black tipped. They range just below the low tide line from North Carolina to the Gulf coast of Florida.

The shells, white shrimp and stone crab were painted at the Duke Marine Lab when we were kindly allowed to use space in the lab of Dr. Bill Kirby-Smith.

Oysters are truly prolific. A single female can produce up to 100 million eggs annually. The larvae are microscopic and go through a series of planktonic stages where a variety of physical and biological forces attend dispersal, aggregation, development, and settlement. Oysters don't live solitary lives. Unarguably, there is only one individual per shell, but where there is one shell there are always more, in most cases, many more. Indeed, very large structures that consist essentially entirely of oyster shells many feet deep can develop. It provides an environment created, maintained, and modified by the animals themselves. Animals end up settling in large reefs, attached to the shells of their predecessors. Competition for space among individuals is intense but beds can be extensive and hold thousands of individuals. Oyster reefs also afford space and protection to other species. Because of their exposure to tides and currents, oysters are dependent on holdfasts and are excluded from soft bottoms. Barnacles, sea squirts, sponges and bryozoans live peacefully attached to the exterior of the shells.

Oysters have been eaten by humans from time immemorial, as evidenced by their presence in Amerindian middens. The extent of these deposits gives some indication of their abundance. An oyster bed can produce a complex habitat that exceeds the sum of a simple collection of individuals. In aggregate, the shells of hundreds of individuals accumulated over decades, provide living conditions for a diverse collection of other settling organisms, small fish, seaweeds, and predators. As bed size increases and the architecture modifies flow patterns of the currents, with their food particles and dissolved gases, the local ecology can change. Abandoned or unused shells provide microhabitats for other species. Oysters provide sustenance to sea stars, a variety of gastropods, and crabs.

Stone crabs (*Menippe mercenaria*, Plate 22, pg. 119), can eat up to 200 oysters each year. Flatworms, fishes (drums and rays), and birds including oystercatchers, scoters and other sea ducks, take oysters. Sponges, boring clams, mud worms, and crabs are commensals that can cause harm. The biological activities which modify current patterns and access to food in turn will influence the occurrence of epibionts, those species that live on the oyster-shell substrate. Epibionts, second-level colonizers, may influence individual growth rates, especially in filter-feeding species. Epibiontic seaweeds (macroalgae) attached to an individual in an oyster bed may reduce the energy of the water flow, enhance sedimentation deposition, and ultimately bury – and suffocate – the animal. All of this is part of the sea's natural food web. As a result of the patterns of settling behavior, oysters engineer large beds, which are often exposed to rather physically demanding conditions, in shallow water.

The loss of economically important oyster beds in recent times off the waters of New York, New Jersey, and Connecticut was due to pollution and mismanagement, not natural predation. The pollution, especially land runoff and silting, has been rectified somewhat by environmental cleanup. Oyster farming has enhanced production for human consumption, but in turn is also susceptible to climate, chemical pollution, and disease.

In addition to local consumption, oysters have been exported overseas and introduced around the United States. Oysters were shipped to France in the 1860s and England shortly thereafter. They went originally as food, and the half-dozen different attempts to introduce commercial beds were unsuccessful. Attempts to introduce oysters to the west coast of North American met with only limited success and breeding populations died out almost everywhere except Hawaii.

The reasons for this are unclear as populations have been introduced successfully within the natural range and continue to be managed to optimize production. A positive side-effect is that restoring oysters can improve the water quality of the estuarine habitat.

Of Shells

Catesby wrote of shells:

> "Every Species of Shell-fish inhabit particular Parts of the Sea agreeable to their Natures: This seems to have some Analogy to Plants, whose different Kinds affect a different Soil and Aspect" (I-xliii).

Marine mollusks make huge quantities of shell. One estimate places the rate of calcium carbonate production by most mollusk species in general as up to 3 lbs. per square yard annually and oysters, specifically, up to 189 lbs. per square yard annually. Compared with the production of wood in temperate forests of about 30 pounds per square yard per year, this means that mollusks produce structural material at rates that may exceed trees! Shells, because of their insolubility, persist over long periods of time. Their production is ubiquitous and because of their durability and density, shells and shell parts accumulate on both spatial and temporal scales. Shells in aggregate provide substrate for subsequent settling organisms, and an internal space protected from the currents, mechanical insults, and predators. Together, they influence the transport of particles and solutes in the immediate environment.

Coral heads, large oyster beds, and mussel-encrusted rocks are immovable objects; just ask any boater who has run aground on one. Despite their appearance as massive and permanent structures they are dynamic systems. The nature of the surrounding water will determine what species come and go, their subsequent success, and the health of the entire structure. In at least one case, the decline in the health of a distant coral reef, a refugee predator species moved to a species-rich oyster reef. The result was an abnormal concentration of refugee organisms that had a destructive impact on the local populations.

The launch of the Industrial Revolution brought with it increased levels of atmospheric carbon dioxide (CO_2). Carbon dioxide is a major end-product of the burning or oxidation of organic fossil fuels, coal and petroleum. It is also the major end product of animal respiration, produced when fuels such as carbohydrates are metabolized in the body. Carbon dioxide is a tasteless, invisible gas spewed directly into the atmosphere. In its frozen state we call it "dry ice" and it is never found as a liquid. Carbon dioxide is considered a greenhouse gas because it prevents re-radiation of solar heat, trapping it in the atmosphere in much the same way as glass works in a greenhouse. The mass of atmospheric carbon dioxide has increased more than 70% in the past 200 years mostly as a consequence of human activity, leading to an increase in worldwide air temperatures.

On a global scale there have been profound effects beyond the temperature increase. The fate of human created carbon dioxide is not limited to the atmosphere. Recent measurements indicate that almost half (48%) of the carbon dioxide from fossil fuel combustion and the production of cement is dissolved in the oceans. At normal atmospheric pressures carbon dioxide dissolves rapidly in water or body fluids. Higher pressure will force more into solution, a process used to produce carbonated drinks such as soda and beer. In our body under physiological conditions, the enzyme carbonic anhydrase combines carbon dioxide with water (H_2O) to form carbonic acid (H_2CO_3). Carbonic acid dissociates spontaneously to bicarbonate (HCO_3^-) and a hydrogen ion (H^+). Hydrogen ion is characteristic of acid. Sodium bicarbonate is a common kitchen product. In the oceans carbon dioxide dissolves readily at the surface and its abundance increases with depth because of the increased pressure. All this activity involving dissolved carbon dioxide levels at the ocean's surface produces carbonate ions (CO_3^{-2}) and hydrogen ions. The result has been an increase in the ocean surface acidity by 0.4 pH units, a rather dramatic change.

There are biological consequences of these simple, but massive, chemical changes. Carbonic ions (CO_3^{-2}) react with calcium (Ca^{+2}) to form calcium carbonate ($CaCO_3$). Calcium is readily available as it is the most abundant divalent ion in the sea. Not insignificantly, calcium carbonate is the main constituent in mollusk shells, corals, and many other marine organisms. The pathway of carbon dioxide in the ocean involves physical features, biogenetic interactions, and is a part of a dynamic system of movement of carbon through the environment. Shells are normally insoluble, but the increased acidity will lower the available calcium.

When the animals die and the shells sink into water under higher pressure and colder temperatures (both of which encourage carbon dioxide solubility) up to 65% of the skeletal calcium carbonate is naturally redissolved and recycled. One consequence is the incorporation of calcium carbonate into shells, which removes carbon dioxide from the air, also making the water more acid. The first is good, the second is not. This is chemically the reverse of putting baking soda (bicarbonate) on a vinegar spill and watching it bubble carbon dioxide. In the ocean, the increased saturation of carbon dioxide can induce acidification, which affects shells. As the shells settle in the water column or are moved away by currents, they are dissolved over a wider range of depths. Paradoxically, as this water is returned to the surface it takes up more carbon dioxide from the air. While this helps offset the increases in human-created carbon dioxide, it inhibits the production of calcium carbonate available for shell formation. While the transfer is good for the atmosphere as it reduces a greenhouse gas, it is not as fine for shelled animals in the oceans, as it interferes with growth. It is currently unknown how the changes in production of corals, bivalves and calcareous plants might influence other species such as those relying on plankton for food. But the consequences of the annual transfer of 118 billion metric tons of carbon from the atmosphere to the oceans cannot be taken lightly.

The problem might not end with plankton and mollusks. Rates of calcification of coral reefs worldwide are reduced compared with those of only a few decades ago. Bird bones and eggshells also are constructed of calcium carbonate. In laying birds, if dietary levels are inadequate, calcium is drawn from the skeletal reserves potentially weakening bones of the laying parent. In penguins, which require very thick shells because their nests are essentially unpadded rock scrapings, and the shell supports the weight of the incubating parent, dietary calcium carbonate can affect breeding success. A weakened eggshell can break and the chick will be lost. It is now known that penguins in Patagonia will increase mollusks in their diet during the breeding season to accommodate the calcium deposited in egg shells. The stomach acid solubilizes the ingested shell and the calcium is absorbed, incorporated into the eggshell, and strength is increased. Humans, too, use dietary calcium supplements to strengthen bones or moderate various bone diseases. Still, it is frightening to think that the calcium dioxide produced in an auto exhaust or from a factory burning coal to generate electricity may eventually reduce the nesting success of penguins in Patagonia.

Engineered habitats are not immune to human-induced changes. Oysters and mussels historically have been over-exploited for food, had their habitats polluted, or reduced mechanically, and suffered invasion from exotic species. These resource-dependent structures have also been affected from a distance by human-caused environmental changes such as erosion, enriched nitrogen content from the use of fertilizers, and changes in the patterns of current flow as the result of construction. Consequently, especially in the western Atlantic these systems have occasionally collapsed. Collapse may follow the removal of a keystone species, an outbreak of disease (often related to temperature change), the arrival of a new predator, or direct physical destruction by dredging or trawling.

Oysters in the Chesapeake continue to decline largely due to decades of heavy harvesting, disease, and numerous environmental pressures. Ironically, about a decade ago, resource managers, industrial interests, and policy makers suggested the introduction of *Crassostrea ariakensis*, the alien Suminoe oyster. This non-native species could potentially rapidly populate the Chesapeake to provide relief to the failing oyster industry, reduce algae levels, and consequently improve water quality. Stakes are high and the debate over the introduction of an alien species accordingly intense. The problem is not settled. Commercial fisherman, government agencies at several levels, resource managers, scientists, politicians, and environmentalists all have interests and opinions.

In 2009, by an Executive Order, President Obama called for a large-scale, "tributary-based", oyster restoration project in the Chesapeake. The resultant agreement involved the National Oceanic and Atmospheric Administration (NOAA) partnered with various federal, state and local agencies to restore oysters in important tributaries of the Bay. The goal is to restore at least 950 acres along the eastern shore by 2025. This broad approach considers salinity levels, available bottom for restoration, and protection from harvest, natural recruitment patterns, and other physical and biological criteria. Cooperating organizations include the Maryland Interagency Workgroup, Department of Natural Resources, US Army Corps of Engineers, and various academic groups and corps of the Chesapeake Bay Foundation with plans to implement, monitor and track progress of the sites. Because multiple locations are involved and resources such as hatchery seeds are limited, the plan is complex. However, necessary survey work is well underway and early construction has begun in several locations -- a significant start in the restoration of a lively community.

"*Mokita*", our Cape Dory 330 cutter

V

OCEANS FORMERLY FULL OF FISH

Fishin' For A Livin'

The Ottis' Fish Market sits on the waterfront in Moorhead City, North Carolina. The eponymous restaurant is next door. An informal color-coding distinguishes the two units: a bright yellow eatery with dark blue trim and awnings and a washed-out, sea-foam-green market. The market is actually a large shed that seems to have expanded from the street front to the pier. It backs directly on the waterfront, which faces the east end of Bogue Sound. Further on the State Docks, a highway bridge, and the turning basin are visible. Across Bogue Sound are Atlantic City and the Coast Guard Station.

Snappers, Bass and Graysby

On the docks, we watch the crews off-load and sort their catch. Boats spend several days at sea and life aboard is demanding and sometimes dangerous. Fishing is hard work, accommodations minimal, and the food on a small commercial trawler may not always be gourmet. Meanwhile, a half-dozen brown pelicans (*Pelecanus occidentalis*) swim close by and several gulls occupy the nearby pilings. Their interest in the various bits and pieces is more than casual as they, too, benefit from the harvest. The captains and crews are locals, often third generation at sea. They gripe about the quantity of the catch, the weather, and the state and future of the fisheries. Their boats demand constant attention, the weather can be miserable, fuel is increasingly costly, and no one can predict reliably what the catch will be. The catch reflects seasonal movements of the fish, and as the seasons progress, certain species are closed for protection and size limits imposed. "It's not like it was for my daddy, or when I was a kid." The litany is repeated in fishing ports along the Chesapeake and through Gloucester to Maine. After a brief visit with family, they will be at sea again. It's all part of the rhythm of "fishin' for a livin'".

On the dock, we spoke to Donald, whose job involved brokering the catch of local fisherman to various commercial and retail interests. When the catch is good, trucks from Baltimore, Washington and even as far away as Montreal and Toronto come for the finfish, crab, clams, and oysters. In the large wooden shed with its high ceilings and ample open floor space buyers pace and examine the catch. An industrial-strength ice machine occupies a room of its own. A walk-in cold room the size of a house trailer sits to the left. Running water and large marble-top tables stand near the scales. Fish are loaded into large plastic containers, separated first by species and then by size. Depending on where the fish were taken two or three species may predominate. On the shed floor there may be baskets of small, medium and large red snapper (*Lutjanus campechanus*). This boat's catch includes a few specimens of the beautiful and much less abundant silk snapper (*Lutjanus vivanus*, Plate 23, pg. 129).

Of special interest to us was the appearance of an occasional individual of an uncommon species. Each "one-off" is of interest because they represent some minor part of the fauna, perhaps a remnant of some once-common species, or is of interest biologically but not commercially. The great hogfish (*Lachnolaimus maximus*) was MJ's favorite Catesby fish painting. After an extensive search we found a specimen in Singleton's Seafood Shack in Mayport, Florida.

It's not obvious from the species off-loaded from the boat, but many fish caught are discarded at sea. This loss, called the by-catch, can amount to over half of the total landed. The species may have little commercial value, may not meet legal requirements of size or age, or be a state or federally protected species. Of the fish thrown back some unknown quantity actually survive. The exact numbers are hard to know but directly influence estimates of the actual populations or biomass.

Part of the by-catch may be good eating, and by law the captain is allowed to keep an individual or two for personal use. On this particular day the captain chose a graysby (*Cephalopholis cruentata*, Plate 24, pg. 131), a species native to the area, but now in numbers too low to be of commercial value. These specimens are always of interest to the artist. Donald was especially helpful in identifying all species. He walked through the displays of the catch with an enumerator from the North Carolina Division of Marine Fisheries (DMF) who checked the stock and recorded where the ship had fished. Once a fisherman himself, the enumerator has worked hard to gain the trust of the crews, but has no power to enforce regulations. The DMF is involved with management issues and the work is of potential mutual benefit to the entire industry. Monitoring the catch for both total yield and demographics, that is, the age and size structure of the population can help maintain the long-term health of the fisheries.

A dark shadow haunts this scene. The reality is that the numbers of fish have declined precipitously everywhere. By regulation, boats are limited to the types of gear they can use and where they are allowed to fish. In many areas the numbers of days spent at sea are restricted; there are catch limits on many species, and operating costs are rising. Fishing, which was once an important source of protein and a profitable industry, is failing. Harvests, even under current restrictions, are not sustainable. Fishermen are seen as both villain and victims. They argue that they need to increase their catch to survive, but the regulations, intended to protect stocks from overfishing, seem endless and draconian. The dilemma is that increased take will further depress stocks and can lead to more global effects on marine ecosystems. Regulations intended to limit catches and to protect the ecosystem cause financial hardship. The balance among resource conservation, the survival of an industry, and the future of an important human food source is at stake. Human history and natural history form a complex tangle not easily teased apart.

Ottis' Fish Market, the seafood business and restaurant, closed in 2003. What had started as a Coca-Cola bottling plant in 1912, became a victim of the reduction in catch and an increase in regulations. One confounding set of factors was regulations originating with the Coastal Resources Commission that designated "urban waterfronts" that made it difficult to run a wholesale business in a public area.

In another local market, ice cabinets display black sea bass (*Centropristis striata*, Plate 24, pg. 131). Along with the closely related groupers, they are tasty table-fish. Black sea bass are slow-growing bottom dwellers, biological characteristics that influence their management, but not their popularity. Most fish in the market are about three years old, as determined from body length. In the Chesapeake, black sea bass are a summer resident and migrate off shore in the autumn. During this movement they are popular with anglers in the Bay. Off Cape Hatteras and in the Mid-Atlantic Bight, their movements are more limited and diffuse and fish are available year round. Adults prefer to live on reefs and rocky outcroppings, occurring closer inshore than the more tropical reef inhabitants such as snappers, groupers and grunts. Because sea bass are active visual hunters, the requirement for light limits their depth. Adults prefer to feed at 20-60 feet. Crabs, mussels, razor clams and other fish provide the bulk of their diet.

The annual total catch of black sea bass in the Carolinas from all sources exceeded 2 million pounds almost every year from 1979 to 1990 and has been relatively constant since then. In the Mid-Atlantic States, recreational landings usually dominated. The catch figures give fisheries experts only a small, but potentially informative, window into the size of the entire population. Based on growth patterns, timing of breeding and morphology, biologists suspect there are two populations of black sea bass in the Atlantic shelf waters, one distributed north of Cape Hatteras and the second from the Cape south. While there is no genetic documentation for this, such information might influence how stocks are assessed. This, in turn, will affect management decisions. Presently, statistics like total catch or numbers of older fish caught are compatible with a healthy population. Fisheries managers believe there is no immediate concern with overfishing this particular species.

Healthy populations are not the case for many other species. Groupers and a variety of species of snappers are heavily fished. These popular species have an extended reproductive season and slow, variable growth rates, and because of high harvests their populations are in precipitous decline. A clearer picture of the factors that influence the size and extent of the resource begins to emerge when these natural history traits are combined with information on primary production and the transfer of energy through the population along the food chain. This ecosystem approach has influenced decisions regarding the protection and conservation of entire fisheries. Unfortunately, even under current regulations, fish populations continue to plummet. Existing allocation of resources has not prevented the vicious cycle that produces serious insults to many species. As concern mounts over stock levels, governments become increasingly involved in issues of management and conservation of the common resources, those that no one owns. The debate rages in all managed fisheries as to the balance of the role of government and the livelihood of fishermen. Ironically, in many countries more money is spent on fisheries management than is earned by the fishermen themselves.

Silk Snapper
(*Lutjanus vivanus*)

The silk snapper, also called yellow-eyed snapper, or Bermuda snapper, can reach 33 in. in length. The yellow iris separates it from its close relative, the red snapper (*Lutjanus campechanus*). It has a long, spiny dorsal fin, and the edge of the caudal fin is blackish. The back and upper sides are pink with a silvery ventral sheen. The sides have fine undulating yellow lines. The fins are reddish or yellow. Silk snappers are abundant off-shore over rocky ledges in deep water along the continental shelf from North Carolina and Bermuda to Brazil.

The lane snapper (*Lutjanus synagris*) that Catesby illustrated (I-17) is found in-shore, over sandy bottoms and would have been easier to obtain than the deep water silk snapper.

This was the first fish I illustrated for this book, obtained from a fish market in Beaufort, North Carolina. We cooked it in butter and white wine. It was delicious.

Larvae stage

— MJ Brush

Lutjanus vivanus - Yellow eyed snapper
Beaufort NC 2003

Black Sea Bass
(*Centropristis striata*)

The black sea bass, also called rock bass, grows to 26 in. Body color may vary from gray to brown to blue-black, and fades to a paler color underside. It is a stout fish with a large head, oblique mouth and large eyes set high on the skull. The dorsal fin is long, the pectoral fins long and rounded. The scales are silvery, large, and patterned, forming longitudinal stripes along the back and sides. Considered a warm-water fish, it ranges along the eastern seaboard from Canada to the Gulf of Mexico. It is common in shallow water over rocky bottoms and rock jetties. It has been found to depths of 425 ft.

This was the second fish I painted while at the Duke Marine Lab. Unfortunately for me, it took so long to paint that instead of becoming dinner, I returned it to the sea.

Graysby
(*Cephalophois cruentata*)

The graysby, also called coney Barbados, has a maximum length of 16 inches. It is a plump fish with rounded fins and a large grouper-like mouth. It is light brown to grey in color, covered with many small brown or red spots and noticeable dark spots under the dorsal fin. The graysby ranges from North Carolina to Bermuda and south to Trinidad, and is found grazing above coral reefs and rocky ledges.

Catesby illustrated the yellow-finned grouper, "*Perca marina venenosa punctata*". He wrote:

> "This Fish has the worst Character for its poisonous Quality of any other among the *Bahama Islands*, but whether they are eatable from any particular Places I know not; many of their poisonous Fishes being not so when caught in some Places; of which the Inhabitants can give a near Guess, but sometimes they are miserably deceived" (II-5).

It is now known that what he described was ciguatera poisoning; an illness caused by eating fish that contain toxins produced by the marine microalga *Gambierdiscus toxicus*. People who have ingested ciguatera may experience nausea, vomiting, and neurologic symptoms such as tingling fingers or toes. The toxin is found in large reef fish.

America's Fisheries

Catesby illustrated 46 fish species on 31 plates. Most are fresh-water species. Catesby's text was skimpy, but he noted large river-runs of herring, which would almost certainly include shads and sardines, and the commercially important menhaden (*Brevoortia tryannus*). Although the illustrations were not considered as accurate or informative as those of the birds, many became the source for information on North American species used by subsequent European authors. Catesby's fish illustrations were described scientifically by David Starr Jordan in an 1884 publication of the *Proceedings of the US National Museum*. Jordan (1851-1931) was an ichthyologist who held both MD. and Ph.D degrees. In addition to being President of Indiana University, and the first President of Stanford University, he was a vocal peace activist in World War I and an expert witness at the 1925 Scopes "Monkey" trial in Tennessee.

Barely 150 years after Catesby, the fisheries along the North Atlantic coast had begun to decline. Commercially valuable fish stocks were being depleted, especially near large population centers. Congress, prior to taking any action, desired documentation of the reports and created a commission to investigate the state of American fisheries.

Spencer Fullerton Baird (1823-1887), then with the Smithsonian Institution, was appointed by President Ulysses S. Grant to be the first Commissioner of Fisheries. A wise move indeed as Baird was a far-sighted leader of the Washington scientific community. In addition to the original charge, Baird even pondered whether the oceanic fish populations might be increased by aquaculture techniques.

The fisheries of immediate interest were the perch-like scup (*Stenotomus chrysops*), a popular but over-exploited species; menhaden, a species not eaten but valued for its oils and use as fish meal; and the anadromous American shad (*Alosa sapidissima*) and Atlantic salmon (*Salmo salar*). Dory fisherman argued that a few entrepreneurs who placed fish traps and weirs in streams and coastal waters were responsible for depleting the populations of shad and salmon which migrated from the sea to fresh-water breeding sites and then returned to the sea. Commercial fisherman also took a disproportionately larger part of the catch. The problem had become more convoluted as several New England states, particularly Connecticut with its major shad fishery, had passed legislation that was intended to eliminate fish traps in their waters. While there was no direct evidence that populations were actually depleted or that traps were the cause, Baird encouraged states to establish other restrictive legislation, such as restricting the use of traps to alternate days. However, no uniform act was obtained, as states did or did not cooperate. Enforcement was difficult at best, similar in some ways to the problem of policing prohibition. Restrictions were quickly repealed but the tension of setting catch levels exists even today, with all the attendant political and economic problems.

In the last quarter of the nineteenth century the United States Congress directly funded scientific work and exploration. Besides Baird's activities at the Smithsonian, Congress supported the work of such luminaries as O. C. Marsh (1831-1877) in paleontology and John Wesley Powell (1834-1902) for exploration and geology. Baird also encouraged governmental involvement in fish culture which had recently been introduced by private interests, mainly to raise trout for sport. In 1873, Congress funded Baird's efforts to raise salmon, shad and striped bass (*Morone saxatilis*) for release in eastern rivers. He kept and bred fish in ponds on the Mall near the Washington Monument and found them to be a cheap and abundant source of protein. The project also included whitefish (*Coregornus clupeaformis*) for the Great Lakes and Atlantic cod (*Gadus morhua*) and halibut (*Hippoglossus* spp.) in the ocean. Although these efforts mostly failed, Baird did manage to ship several tank-car loads of young shad and striped bass west for release in the Pacific. Both are established there today and have some commercial value. This is not unlike the later attempts to introduce the Eastern oyster to the Pacific shores, which have also met with limited success. Baird managed successfully to import and establish carp (*Cyprinus carpio*). The fish were widely distributed and carp are now plentiful in inland waterways across North America. This is an example of a successful alien species introduction. Unfortunately, carp are considered a pest by freshwater sports fisherman: another example of good intentions gone awry.

Under Baird's leadership, a laboratory for the study of fisheries, fish-culture and marine sciences was established permanently at Woods Hole, Massachusetts. The Woods Hole facility became a year-round facility by 1883 when Baird arranged for land, funding from private sources such as Harvard University's Alexander Agassiz, and additional construction funds from the Congress. Less than five years later a second laboratory was authorized by President Theodore Roosevelt to operate at Beaufort, North Carolina. Woods Hole functioned as a research laboratory from the start. Investigators from Johns Hopkins, Harvard and Princeton Universities were among the first involved. Early work targeted Vineyard Sound's mollusk beds to see if depletion of this food source was somehow related to the drop in coastal fish populations.

The Woods Hole laboratory originally supported summer research and teaching. Baird's successor as Commissioner of Fisheries, George Brown Goode (1851-1896), expanded the Fisheries Commission mandate to include scientific research (quaintly called "Inquiry" at the time), fish culture and statistical studies.

Further, Baird attempted to find new fishing grounds for American fishermen, and to find new marketable fish species. One possibility, the perch-like tilefish (*Lopholatilus chamaeleonticeps*) described scientifically by Goode in 1879, unaccountably disappeared from where they were first found in relative abundance and never became reliably harvestable. It is now known that this species is sensitive to cold water and a mass die-off occurred in 1882. The species remained rare for decades presumably because of the cold water in the area, as tilefish were probably never over-exploited, and a small long-line fishery exists today. Physical changes such as these are now known to accompany large climate shifts such as the North Atlantic Oscillation. In April 2003, 700 tons of Atlantic cod froze to death off Newfoundland in a similar incident. The Woods Hole facility was one of Baird's great legacies. His roles as an organizer and promoter of science at the Fish Commission, at the Smithsonian, and in the National Academy of Science were unparalleled. The laboratories are now home-ported in the Department of Commerce within the National Oceanic and Atmospheric Administration (NOAA) and the Bureau of Commercial Fisheries, now the National Marine Fisheries Service, was incorporated from the Department of the Interior. The contributions of these laboratories have been enormous, but they still struggle to establish completely successful conservation and management programs.

Chasing A Diminishing Resource

Fishing in the world's oceans was generally unregulated through the first half of the twentieth century. The traditional three-mile limit that established national waters was framed in 1822, and defined the waters over which a country maintained jurisdiction. Little attention was paid to the open ocean, as the riches there were considered inexhaustible. Local control in many places was lax, and the fisheries in many in-shore fisheries locations were depleted by the mid-eighteenth century. Cod fisheries were notably depleted in Great Britain by the early 1900s and in Iceland scarcely three decades later. These changes were driven by improvements in technology, increased numbers of boats fishing, and in part by an increased demand for cod-liver oil as a source of vitamin D (which did much to eliminate rickets in these countries). The Grand Banks, a common North Atlantic resource fished by several countries, was also depleted of cod.

During World War II fishing in the North Atlantic was greatly curtailed. Consequently the Grand Banks, North Sea, and Baltic Sea fish populations, especially cod, rebounded from the earlier depletion. However, after the war, as navies of the world provided ships easily converted to fishing, fish became an affordable source of protein, and fish products had vast commercial potential. When the catches began to fall, policies regarding access and limits were instituted. These changes in territorial claims challenged the notion that nations had free access to the world's oceans. It was no longer an unrestricted resource, freely exploitable. By 1945, the international agreements regarding access to the sea were challenged for a somewhat different reason. President Harry Truman proclaimed the United States right to control mineral resources on the continental shelf. Previously no one – no state, no country – owned its continental shelf. Truman was primarily concerned with oil and gas but the implications for fisheries were immediately apparent. What was previously treated as a common resource now became a national property. Accordingly, Truman also declared conservation zones in the areas of the seas contiguous to the coast to conserve and protect fish.

Truman's actions set off a series of debates, claims and counter-claims about the propriety of the actions and the consequences to various countries. The nations of Western Europe, Russia, Canada and the US, plus all the newly nationalistic countries of South America, were involved. The resulting battle over access to fisheries continued for over 20 years. Countries sequentially extended their claims: for example Iceland, a relatively small island nation with an economy almost totally dependent on fishing. The economic consequences were immense. A fracas broke out over cod and herring that involved the United Kingdom and Scandinavian countries. This lead to a series of encounters at sea, often called the "Cod Wars". Historically, this was nothing new. In the 1700s England and France had battled over fishing rights in what became Canada. Various diplomatic attempts at reaching compromise over fishing rights for cod involved the International Court of Justice, the United Nations and North Atlantic Treaty Organization. Eventually the European Economic Community declared a 200-mile exclusion zone. By mid-1970, the 200-mile zone was essentially a worldwide standard, which functioned basically to exclude foreigners from a nation's fishing grounds. Island nations had exclusion zones, often larger in area than the islands themselves. What was required next was for each country to regulate the catch within its zone, something much more difficult. With sustainability as a goal, each country became responsible for management of its own resources. Government regulations became the tool to control harvest, although one country could allow another to fish its territories for a fee. In some places citizens were no longer free to fish when and where they wanted. Despite these measures, very few fisheries have recovered and none to their previous levels. The consequences of these actions are very much with us today. Nations realize they must regulate both the kinds and amounts of fish taken.

Here's how the game gets played. Fishermen want a reasonable yield and a profit for their labor. Better fishing technology makes for better yields. Governments realize they must prevent a valuable resource from being destroyed. Good management procedures imply that populations must not be over-exploited at the risk of a crash. So regulations are imposed. For example, larger mesh trawler nets were introduced to reduce the numbers of smaller fish in the catch. Presumably these individuals would grow to breed and provide catch for the future. The unintended consequence was that with the depletion of the larger breeders, the younger, smaller fish began to breed. The predictable results: artificial selection for smaller individuals. It follows that the net yield is reduced so perhaps more boats, often financed with government subsidies, would come into service to catch more of the smaller fish. Fleet size is increased, to chase a contracting population. In response, government may limit the number of days the boats can spend at sea. Fishermen develop larger, more efficient gear to be used fewer days at sea. The catch is maximized, but stocks continue to decline. The government may put quotas on the species taken by each vessel per season. If the wrong species – one for example that exceeds the species limit – is hauled up from depth the pressure change may kill it. It is waste and adds to the by-catch. Governments in some countries have then responded by closing fisheries and relocating the displaced fisherman. In other areas, reserves, or "'no take zones," have been established. Both have met with limited success, as the fish numbers often increase, but the physical damage to the bottom from trawling may take generations to repair. By the same token some populations may recover, perhaps enough to allow minimal harvest, but never to reach a large enough breeding population to restore original numbers.

"Who Will Get The Fishies When The Boat Comes In?"

The interactions among species often hold surprises, as shown by a recent comprehensive study of large sharks. Population studies along the entire Eastern seaboard and long term tagging studies off the North Carolina coast have shown declines for large shark species significant enough to deem them essentially eliminated from the ecosystem. The species involved plummeted because of the intense demand for shark meat, by mortality from being by-catch in many fisheries, and from sport fishing. These apex predators feed on rays, skates, and smaller sharks. With release from predation, twelve of the 14 prey species underwent population increases. The effect of this restructuring in the community was that predation on bay-scallops by the smaller species (meso-predators) became sufficient to destroy the century-long Carolina Bay scallop fishery. Cascade effects such as this illustrate the care that must be taken in managing marine ecosystems.

Governmental attempts at regulation, while well-intentioned, have been inadequate to maintain the worldwide populations of fish. The industry became more efficient after World War II with the use of SONAR, electronic navigation, stronger nets, more powerful winches and other advances. The industry struggled with a declining resource and in some areas overwhelmed the regional fisheries. The economic consequences were severe. A classic example is the North Atlantic cod fisheries. Over-fishing eventually led to devastating social and economic events in Canada and Europe. The Canadian government in 1992 was forced to close much of the Grand Banks and the Gulf of St. Lawrence to ground fishing. The results were both economic and social upheaval. Globally, fishing has expanded to the point where effectively all fisheries are under heavy stress and several species, while they haven't disappeared completely, are essentially extinct, a condition described as "ecological extinction".

Fisheries management focuses on describing demographics, estimating the numbers of fish, and establishing the regulations thought to be necessary to maintain an important resource. Crudely estimable numbers account for the fish taken by many species of birds, other fish species, disease and natural disasters, all of which reduce populations. For many species, the numbers taken by humans have not been sustainable and yields in many fisheries have fallen dramatically. The decrease in the past 60 years has been greater and probably happened more rapidly than the original conditions that lead to the creation of the Fisheries Commission.

It is clear that the oceans' phytoplankton production can support an immense amount of animal life, more than adequate to support its fish populations. Hence, from a fisheries perspective, the problem is not with inadequate basic production, but rather the effectiveness with which the plant material is converted into fish. It takes huge quantities of microscopic phytoplankton to make only a small serving of top predators like tuna or swordfish. The movement of energy itself through the food chain is complex and better described as a food web. Because only the larger fish are ultimately harvested, we sample the highest, most expensive levels in the hierarchy. Further, there is a significant loss, more precisely wastage, in the by-catch. Unwanted, or illegal, fish, mammals, birds, turtles, and invertebrates are all caught incidentally to the targeted species, a further loss of resources.

For historical reasons, no accurate pre-exploitation baseline values for the world's fish ever existed. Now, understanding of the health of fish populations worldwide is dependent on accurate reporting by each country. The best estimates, those by organizations such as United Nations' Food and Agriculture Organization (FAO), indicate that the size of fish catches has declined significantly, some say alarmingly, since just 1988. The large predators that remain, including sharks, tuna (*Thunnus albacares*, Plate 25, pg. 139), marlin and swordfish, are scarcer and individuals smaller than just 50 years ago. Fishing boats are larger, and range further. As the catch falls in one area the boats move to another, or shift to a different species or use trawls rather than long-lines or nets. The first area may be left to recover, but there is no way to predict how rapidly or to what extent recovery will occur. Recovery is based on the number of breeding fish that remain and the rate the population can replace itself; but, as we know, recruitment, population size and productivity are all difficult to assess.

Fish provide less than 20% of the world's annual protein intake, yet this is an amount too important to be overlooked. Commercial fishing, which accounts for about 80 million tons annually, is not the only draw on this resource. About 300 species of seabirds, roughly 700 million individuals, take about 70 million tons of fish annually. Recreational fishing at sea takes a significant portion. If the industrial fisheries for menhaden and pollock, for example, are excluded, recreational fishing accounts for over 20% of the total of seriously over-fished species along the Atlantic coast. Several, such as red drum (*Sciaenops ocellatus*) are "species of concern." Recreational takes rise to 38% in the South Atlantic and 64% in the Gulf of Mexico. The recreational catch for summer flounder (*Paralichthys dentatus*), scup (*Stenotomus chrysops*), and red snapper (*Lutjanus campechanus*) can approach the amounts taken commercially. The initial regulatory reaction to these statistics was to tighten restrictions on recreational takes by lowering limits on given species. These restrictions may be tightened further in the future by the initiation of a lottery system similar to those employed for commercial salmon fishing on the west coast of Canada and Alaska. Management continues on an area-by-area or species-by-species approach, and needs to be expanded.

The size of the remaining stock of a fish species cannot be accurately estimated. Managers have few alternatives. Options include reduction in the number of days of fishing allowed per season, changes in the gear used, or reduction of specific catches. Quotas can be tailored to specific species found in areas at risk. Finally, and perhaps most draconian, would be to impose quotas regardless of fleet size, which would forbid all fishing in an area. Each of these options would eliminate thousands of jobs, millions of dollars in revenue, and significant amounts of protein from the human diet. Unfortunately, there is an ongoing race between regulators whose intent is to restrain fishing and fishermen whose livelihood depends on thinking up ways around them. And estimates of depletion of fish stocks do not include that taken by seabirds and mammals.

An evolving approach to fisheries management worldwide is to ensure that the total biomass removed from a given fishery does not exceed the total amount of the system's productivity, and that the integrity of the ecosystem is maintained. Traditionally, the approach to fisheries management focused on single species. With the collapse of world fisheries, focus shifted to consideration of the impact of harvesting fish on entire ecosystems. The problem of quantifying the effects of human activity on the scale of ecosystems persists, but attention has shifted to more universal considerations such as conservation of sustainable populations, maintenance of biodiversity, and protection from the effects of pollution and habitat destruction. This approach still encompasses the attention to individual growth, population recruitment, recovery and rebuilding times, understanding species interactions and identifying those stocks most threatened by extinction. It is also understood that evolving management practices will incorporate the strengths of various population models, the legal and economic implications of regulation, allowable modifications to gear to control the by-catch, and the role of marine protected areas. International cooperation is crucial to make fisheries reform effective.

Current modeling rests on a firm basis of biogeochemical and oceanographic information. The origin, movements, levels and economy of chemicals in the ocean are well established, as is the nature of the oceans' temperature, currents and other physical features. The challenge is to extend these models through the food web at scales relevant to processes that include the living matter: plants, plankton and especially fish. Organisms at different trophic levels have different life histories and can vary widely in abundance and distribution. Further, only recently has it become possible to incorporate the effects of human-caused habitat destruction, the incidental

Yellow-fin Tuna
(*Thunnus albacares*)

The yellow-fin tuna is a large fish reaching 50 in. in length. Size varies by region. Common names include yellow-finned albacore, Allison's tuna, Pacific long-tail and yellow-fin tunny. The torpedo-shaped body tapers sharply toward the thin tail. The long yellow dorsal and anal fins lengthen with age. There are seven to ten yellow dorsal finlets. The top of the body is a shimmering dark blue to greenish blue and the lower sides are a silvery white crossed with vertical interrupted lines. A golden stripe runs from behind the eye to the tail.

Yellow-fins are distributed worldwide and are highly migratory in tropical and subtropical waters. They travel in large schools and are extremely active and attain speeds up to 50 miles per hour.

The yellow-fin diet includes cephalopods, crustaceans and other fish, including other tunas. They appear to be sight-oriented predators, feeding during daylight.

This watercolor was painted from video and still photographs, primarily from the website: ARKive.org.

mortality of non-targeted species, shifts in population demographics, and the responses in a species when the numbers or availability of another species changes. Not all ecosystems have been studied equally and models cannot be expected to respond in the same ways. Alone and in combination these factors, plus the potential effects of political and economic influences, make management a risky business.

Working at the level of the ecosystem is potentially a fruitful approach, but questions remain. As is the case with all the earth's ecosystems, humans have dominated and changed the oceans. There have been additional detrimental alterations to coastal ecosystems which have an effect. Factors that affect stability include changes in diversity, in the strength of the interactions among species, in the topology of food webs, and differential sensitivities to environmental perturbations. The overall viability of a particular system can be inferred from the rate of return to equilibrium after repeated insults. Consequently, the extensive and often irreversible changes to ecosystems are accelerating and abrupt; many include important consequences for human well-being, including, of course, productivity of sought-after fish.

While establishing marine reserves and closing fisheries may increase local species diversity and productivity, stability and community variability may not recover. There are uncertainties as to which habitat to protect, strong opposition from the fishing industry, and multiple challenges in enforcing even the existing protected areas. Even deep in the oceans, where new levels of biodiversity have been revealed, procedures such as bottom trawling have affected cold-water coral reefs and sea mounts. The problem is accelerating world-wide. Something of a race has taken shape between those exploring new resources using new technologies, and those trying to maintain sustainable harvests. Hanging in the balance is the health, or even the existence, of viable global fish stocks.

VI
BAHAMAS & BEYOND

700 Islands In Some Of The World's Clearest Water

Our first glimpse of Eleuthera, neighboring St. George's Cay, and the town of Spanish Wells, was from the air. Bright blue, green, and turquoise water covers the dazzling sandy flats. Small fishing boats dot the waterscape. Eleuthera is only one of the 700 or so islands that lie on the windward side of the continental shelf and directly in the path of the westerly ocean currents that bring the fertile deep-sea waters so ideal for reef-building organisms. To the leeward, the shallow banks lie in warm, clear seas protected from winds and currents. The islands are isolated from coastal North America by the Gulf Stream, which flows north, is narrow and swift, and has excavated a channel 2600 ft. deep. This group of islands, composed of coral reefs, may have begun to form 350 million years ago.

The Bahamas have been islands for their entire existence. During the last glacial period that ended 8,000 years ago, the northernmost islands that border the shallow banks were probably interconnected. Subsequent changes in sea level submerged the banks and produced the present islands. The larger land mass meant more available habitat and facilitated movement of plants and animals among the islands. The elevation on any of the islands rarely exceeds 15-30 ft.; the highest point in the entire Bahamian chain is Mt. Alverina (200 ft.) on Cat Island. It has been calculated that if sea level rose only 10 ft., more than half the Bahamas' land area would be inundated.

In geological terms the flora and fauna are not very old. The Bahamian avifauna most likely originated in the larger islands of the Greater Antilles including Cuba and Hispaniola and moved from island to island by flying. Lizards most likely rafted on floating debris, aided by current and wind patterns. Within the chain, bird distribution is related to the degree of geographic isolation, the size of the island, the number of available habitats, and the types of vegetation. This ecological variability is affected by factors such as rainfall and prevailing wind direction. In general the Bahamas are neither species rich nor heavily forested.

When Columbus arrived in 1492, the islands were inhabited by Lucayans. Because coral islands lack precious metals and Columbus found no gold, he spent little time there. The Lucayans were related to the Arawak and Tainos, indigenous peoples of Cuba, Jamaica and Hispaniola. All had suffered from the aggressive Caribs who periodically raided the Bahamas and captured men, women and children. Columbus felt the Lucayans would make excellent servants or slaves because of their mild dispositions. He quickly pushed on to the volcanic Hispaniola where the locals were forced to work as slaves in the gold mines. As more labor was needed and the local human population was depleted, the Spanish returned to the Bahamas to enslave the Lucayans. Soon the Lucayans were extirpated and the Bahamas were not settled again until 1640 when a group of Englishmen from Bermuda landed on Eleuthera (Greek for free). Eleuthera is on a fairly direct course south from Bermuda, and the settlers' ship went aground on Devils Backbone, a submerged reef which connects several small islands off North Eleuthera. The water inside the reef is quite shallow and the colonists landed unharmed on a nearby beach. They lost their possessions, but found food and were sheltered in local caves. In a short time a permanent settlement developed on nearby Harbour Island. Dunmore Town today is a site of great charm, typified by its New England style homes painted in brilliant colors. Other islands were subsequently explored and the government eventually moved to New Providence. Prior to the American Revolution, colonials came to the Bahamas for their health. Nassau became the capital and in 1973 the Commonwealth of the Bahamas gained full independence.

Green Sea Turtle

Turtles were an important protein source, often carried aboard ships. Catesby illustrated the loggerhead turtle, (*Caretta caretta*), and the green sea turtle, (*Chelonia mydas*, Plate. 25, p 139). Turtles, especially the green sea turtle, were once extraordinarily abundant in the Caribbean. Prior to Columbus, turtles may have equaled the biomass of all the American bison (*Bison bison*) on the central plains.

Green sea turtles are herbivores, grazing on the abundance of seagrass growing in shallow waters surrounding the Bahamas. "Seagrass" is a useful term to refer to four families of grass-like flowering plants adapted to marine, fully saline environments and widespread in the tropics. They provide stability to the sandy bottoms of lagoons, estuaries, protected embayments, and beaches. Seagrasses moderate wave action, increase water clarity, and provide nursery habitats for many species. They are extraordinary forage for green sea turtles. In fact, an important ecological balance exists between the plant and the turtles. Turtles feed on the meadow-like beds and may establish feeding beds that are used for years. Where turtle populations decline grasses may subsequently die-off for lack of nutrients. Accumulation of waste organic matter, especially sulfates, can increase anoxia, which further inhibits plant growth. Historically, the turtles ate the plants, but deposited the 'fertilizer' elsewhere to moderate chemical buildup. Increased plant density may encourage slime-mold infections, and the mold prefers the older portion of the leaf, that portion previously grazed by the turtles. If the grass remains uncropped, self-shading may develop, not a good thing for a plant dependent on the brilliant sunlight of tropical waters. When seagrass density drops it in turn affects both the physical and the biological environments.

Sea turtles also provided food and materials for humans. In an early attempt to conserve this valuable resource, in 1620 the Bermuda Assembly passed a law to protect young green sea turtles from harvest. Nevertheless, by 1800 the sea turtles of the Caribbean were seriously depleted mostly by over-harvesting for human consumption. Today all sea turtles are listed as threatened or endangered.

Catesby's Bahamas

Catesby stopped in the Bahamas for several months after he left Charleston. Captain General, Governor-in-Chief, Chancellor and Vice-Admiral George Phenney welcomed and assisted Catesby who visited Eleuthera, Andros, and Abaco. The birds of Carolina that had so enthralled him were no longer a distraction and he immediately turned his attention to this new place. Catesby reported "…the greater Number of *Bahamians* content themselves with Fishing, striking of Turtle, hunting *Guanas* (=iguanas), cutting *Brasiletto* Wood, *Ilathera-Bark*, and that of wild Cinamon…" (I-xxxviii). They also raised goats, grew yams, and produced corn-meal bread now known as johnnycakes. Unlike the Caribbean islands, the Bahamas had no sugar crop, hence no rum. Rum and Madeira wines were imported. Madeira wines were known to travel well and became popular in the New World (George Washington is reputed to have drunk a pint daily at dinner).

Like most visitors, Catesby found the Bahamas "…blessed with a most serene Air…". The islands were "…small, having a dry rocky Soyl…void of noxious Exhalations…". Even in the early 1700's "This Healthiness of the Air induces many of the sickly Inhabitants of *Carolina* to retire to them for the Recovery of their Health." The islands had fresh winds, and no snow, frost, or earthquakes. On the other hand, thunder and lightning were frequent and August and September were "blowing Months, and are attended with Hurricanes." Catesby noted that the islands "…may not only be said to be rocky, but are in reality entire Rocks." The soil supports "…a perpetual Verdure, and the Trees and Shrubs grow as close and are as thick cloathed with Leaves, as in the most luxuriant Soil" (I-xxxix). He catalogued the soil and its distribution and remarked on how well it supported the agricultural efforts of the islanders. He also accurately described the fringing reefs and noted: "These rocky Shores must necessarily be a great Impediment to the Navigation of these Islands; but as the inhabitants are well acquainted with the Coasts, and expert in building Sloops and Boats, adapted to the Danger, they do not suffer so much, as the terrible Appearance of the Rocks seem to threaten" (I-xl). Local knowledge is a timeless resource.

Catesby noted only a few kinds of Bahamian land birds but none of the seabirds found in the Bahamas. About 300 bird species have been recorded from the Bahamas and 109 species are reported to breed there because most are vagrants and occur only rarely, or are seasonal transients. During his visit to Andros, Catesby observed bobolinks (*Dolichonyx oryzivorus*) passing overhead at night. He knew the bird as an agricultural pest in the Carolinas that disappeared in the fall. From this he began to formulate an understanding of bird migration. Catesby recorded and named a number of species, none of which is endemic to the Bahamas, that is, restricted exclusively to this group of islands. Indeed, most have a wider distribution in the West Indies, especially the larger islands to the south, a distribution pattern shared with the Lucayans. One species, the Cuban bullfinch (*Melopyrrha nigra*), Catesby's "Little Black Bullfinch," is endemic to Cuba and probably was a cage bird, as Catesby didn't visit Cuba. History does not reveal if the three known Bahamian endemic birds were not seen by him or simply not recognized as different.

Of the seven species of land birds illustrated by Catesby from the Bahamas, most have affinities

GREEN SEA TURTLE
(*Chelonia mydas*)

Adult green sea turtles can measure 4 ft. in length. The oval shell varies in color from light to dark brown; the underside is whitish to pale yellow. The head is light brown with yellow markings.

They are distributed worldwide in tropical waters.

Catesby illustrated the green sea turtle and wrote:

> "THE *Sea-Tortoise* is by our Sailors vulgarly called *Turtle*, whereof there are four distinct kinds: The *green Turtle,* the *Hawks-bill* [*Eratmochelys imbricata*], the *Logger-head-Turtle* [*Caretta caretta*] and the *Trunk-Turtle* [*Dermochelys coriacea*]. They are all eatable; but the Green Turtle is that which all the maritime Inhabitants in *America*, that live between the *Tropicks*, subsist much upon. They much excell the other kinds of Turtle, and are in great Esteem for the wholesome and agreable Food they afford" (II-38).

We observed a pair of copulating green sea turtles in the estuary of Jekyll Island, Georgia, and were able to drift quite close to them without disturbing the pair in the least.

with the Greater Antilles, not the US mainland. In 1974, the Eurasian collared dove (*Streptopelia decaocto*) was introduced into the Bahamas and now occurs on all the major islands. By the 1980s it had spread to Cuba, and is now common on other Caribbean islands. It has also spread to south Florida, expanded north along the southeast coast, and occurs at least as far north as the Cape Fear River. In modern times the smooth-billed ani (*Crotophagea ani*) has become common and is a breeding resident in south Florida. It most likely came from Cuba rather than the Bahamas. Hence, traffic may flow in either direction.

The black-faced grassquit (*Tiaris bicolor*) and Greater Antilles bullfinch (*Loxigilla violacea*) are indigenous, abundant seedeaters that breed throughout the Bahamas. Along with the expansion of open spaces and the increases in grasses and shrubs after the arrival of Europeans, other seed-eating birds were introduced. These include the house sparrow (*Passer domesticus*), Northern bobwhite (*Colinus virginianus*), and Cuban grassquit (*Tiaris canora*), all of which have become established on specific islands. Island groups such as the Bahamas have typically seen changes in the birds associated with the forested lands as they are replaced with more open habitat.

Sour Orange

Sour orange (*Citrus* x *aurantium*, Plate 27, pg. 149) grows everywhere in the Bahamas, but it is an introduced alien. The trees seem to adapt well to the chalky soil. Trees grow to only 10 to 15 ft. tall and an open branch pattern makes the abundant fruits easily accessible. The rough surfaced peel can be bitter and contains cells that produce aromatic oils. The fruit is best squeezed in the skin to produce the lovely pulp-filled juice, which is consumed fresh and perhaps best when mixed with tequila (another product extracted from a plant). It is a favorite ingredient in making English-style marmalade.

Sour orange does well with very little attention and it has a curious history. It originated in Southeast Asia, and there is evidence that it may have accompanied the very earliest settlers on the Pacific islands of Fiji, Guam, and Samoa. By the ninth century it was recorded on the Arabian Peninsula and was grown in Sicily by 1000 AD. By the twelfth century it was cultivated around Seville, Spain. For 500 years it was the only orange grown in Europe. *C.* x *aurantium* was the first orange to reach the New World: Mexico by 1568, Brazil in 1587. Not long afterward it appeared in Jamaica, Puerto Rico, Barbados, and Bermuda. The Spanish, perhaps Ponce de León, introduced Sour Orange in St Augustine, Florida. De Soto planted orange trees across Florida and it was quickly adopted by the early settlers and Native Americans and then exported to England by 1763. Accordingly, it could have come from Bermuda to the Bahamas in 1640 with the first English settlers. It is possible, but considered unlikely, that sour orange arrived with Columbus in 1492.

Sour orange is naturalized in many places in the western hemisphere from Georgia through Mexico to Argentina. There are trees on hummocks in the Everglades associated with sites where Indians once lived. Sweet orange (*C. sinensis*) budwood was grafted on sour orange trees early in colonial times. Even today the sour orange is used as rootstock in many citrus-producing areas worldwide.

Besides marmalade, sour orange has many uses. In Mexico the fruit is cut, salted and eaten with a paste of hot chilies. The juice may replace vinegar in cooking fish and meat. The essential oils, those extracted by cold-pressing or gentle steam distillation, are used to flavor many foods, adding the familiar citrus flavors. Oil from one variety is used in the production of Curaçao, a popular liqueur. Dried flowers are used to add flavor to tea, but orange pekoe tea is not flavored with orange. Orange pekoe refers to the grade of tea, and Lipton's famous orange pekoe is considered a high quality tea. In the Pacific the crushed fruit and macerated leaves will lather in water and are used as soap. The essential oils are indispensable to the perfume and cosmetics industry, and they play a valuable role in aromatherapy.

Caribbean Spiny Lobster

Stand on the beach on any one of the Bahamas' outer islands and look seaward. What you see today is, in essence, no different from what a pre-Columbian Indian, an early English arrival, or Catesby himself would have seen. Brilliantly clear waters, a tropical blue sky, and afternoon fair-weather clouds. No sounds except for the lapping of the very gentle surf and the occasional tern calling. But the view landward is different. In the early 1600s, Spanish conquistadors landed on Eleuthera and sank a well, which provided fresh water for their galleons prior to crossing the Atlantic on the journey home. The first English inhabitants arrived in 1649, followed by Loyalists from Carolina in 1776.

The short ferry ride west from Eleuthera passes a lovely anchorage with several cruising sailboats at anchor. Spanish Wells is a small town on St Georges Cay with an active quayside focused primarily on its spiny lobster fishery. Government dock in Spanish Wells is one center of activity and arriving sailors are cautioned to "Watch for fish scales and slime on the steps." The fish dock is handy and marine services, potable water, and provisions are available. The name Pinder seems to appear on businesses and services everywhere. Most locals, although native-born Bahamians, are descendants of the early English settlers.

Spanish Wells produces three-quarters of all the spiny lobsters taken in the Bahamas. This is not the large clawed Maine lobster (*Homarus americaus*) but the locally abundant Caribbean spiny lobster (*Panulirus argus,* Plate 28, pg. 151). 30 species of the genus *Panulirus* occur worldwide in tropical and temperate waters. They tend to inhabit rocky habitats or reefs, which provide cover and protection. As they occur in shallow waters and may be locally abundant, humans have probably taken them for food since time immemorial. Smaller adult animals live in sheltered bays and estuarine habitats where they are often the top predator. These areas typically feature lots of nooks and crannies among the mangrove roots, rocky outcrops, ledges, corals and sponges. But they do not breed here. As larger adults they move to offshore communities and occupy rock and coral crevices. Site fidelity is high and den sites are selected to maximize concealment and exclude predators. Both sexes may migrate offshore in the fall when the severe storms arrive and inshore water temperatures decrease. Occasionally the autumnal movement includes mass migrations where lobsters march single file connected by their antennules, another aspect of their gregarious nature.

Sour Orange
(*Citrus* x *aurantium*)

Sour orange, also called bitter orange, Seville and bergamot orange, is a globose fruit 2 3/4 in. in diameter with a rough, fairly thick, bright reddish-orange, aromatic bitter peel, that includes sunken oil glands. There are ten to twelve internal segments containing pulp and numerous seeds. The 2 1/2-5 in. long leaves are evergreen, aromatic, and alternate on a broad winged petiole, shiny green above, pale below. The fragrant flowers are1 1/2 in. wide, with five white petals, and arranged singly or in small clusters.

Sour orange is planted in subtropical or near-subtropical climates, but can endure short frosts. It is grown commercially for its health properties and as a flavoring.

I painted this specimen, picked from a tree in Spanish Wells, on St. Georges Cay, an island just west of Eleuthera in the Bahama Islands. Part of the texturing of the skin was achieved by thickly painting watercolor onto the orange and pressing it into the drawing.

We squeezed the juice to make an excellent tequila sunrise.

Caribbean Spiny Lobster
(*Panulirus argus*)

Caribbean spiny lobsters grow to two ft. in length. Common names include the Florida lobster, rock lobster, crawfish and bug. Their body is gray or tan, mottled with shades of green, red, brown, purple or black with large cream-colored spots on the second and sixth segment of the abdomen. They lack the large pinching claws of the North American lobster, (*Homarus americanus*), but a shell covered in spines helps to protect them from predators.

They inhabit the tropical waters of the Atlantic Ocean, Gulf of Mexico, and south to Rio de Janeiro in Brazil.

I painted this specimen from a molted shell I found in Spanish Wells, St. George's Cay, Bahamas. We enjoyed their flavorsome flesh several times in local restaurants, different than the North American species, but equally as delicious.

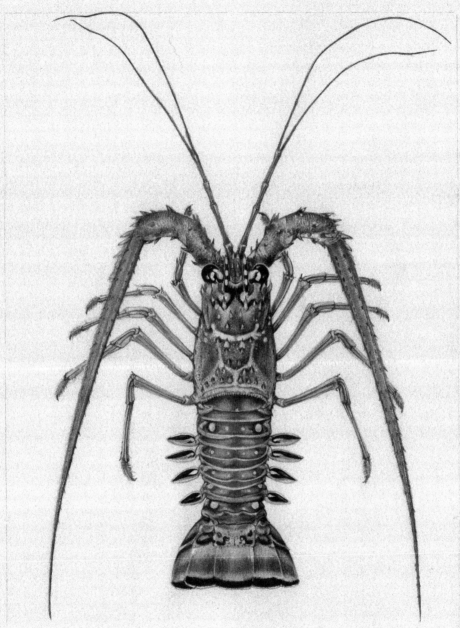

M.J. Brush

4/07   SPINY LOBSTER FLORIDA SPINY LOBSTER   *Panulirus argus*
M.J. BRUSH / MYSTIC CT.   Lots of color variation

Spiny lobster are fished throughout their range and marketed fresh, frozen, and canned. Lobsters are singularly important in the Bahamas where the government has established an exclusive fishery zone. The Department of Fisheries is responsible for its management and development. The shallow waters of both the little and great Bahaman banks yield over 8,000 tons of spiny lobster annually. The eight-month season for spiny lobsters contributes over $60 million to the local economy, making it the most important commercial fishing activity in the Bahamas. In Florida about 11,000 tons are taken annually, with a commercial value of approximately $90 million. Florida also allows recreational fishing for spiny lobster. There is a minimum size regulation in effect, a bag limit, and a license is required. Nevertheless, as is the case with finfish in many areas, lobsters are now breeding at a smaller size than previously. Fishing has caused a shift in the population towards younger, hence smaller, breeding individuals. There are also fisheries for finfish such as the Nassau grouper (*Epinephelus striatus*), conch (*Strombus gigas*), and sponges.

Divers use small (around 18 ft.) dinghies as platforms to search for lobsters either in the natural rock formations or in *casitas*, the artificial habitat made of cinderblocks with tin roofs. Compressors supply air to the diver through a relatively low-tech, low-cost "hookah" system. A larger vessel acts as a base of operations. Commonly, stints at sea run four weeks with the 'mother ship' providing food and sleeping quarters, and storage for the catch. Upon return the catch is sold dockside to the local processing plant. Fishermen have close to 700,000 casitas in place and use over 105,000 lobster traps. The government regulates the catch and the UN Food and Agriculture Organization has assessed its management program positively. An estimated 90% of the catch is exported, much of it to Miami and Fort Lauderdale, where it is loaded aboard the many very large cruise ships plying the Caribbean. The remainder feeds tourists in Nassau, Grand Bahamas and other Bahamian tourist spots. People love lobster.

The elaborate shell of the spiny lobster is actually an external skeleton. Crustaceans lack any hard internal support analogous to our bony skeleton. The spines and other embellishments mainly provide mechanical protection. In addition to sharp spines, the legs (lobsters are decapod crustaceans with five pairs of legs) are jointed like a suit of medieval armor. The carapace is intricately colored with areas that range from light tan, to purple, to a deep red-black. The segmented abdomen, in addition to the various grooves and bumps, sports at least two pair of ocelli. Structurally, ocelli are simple eyes that contain light-sensitive nerve cells, with a lens formed from the outermost cuticular layer. Ocelli can detect the presence of light, but not its direction. Unlike clawed lobsters, spiny lobsters have both antennules and antennae. The antennae exceed the total body length and are highly mobile. Sensory elements associated with stiff bristle-like spines and setae cover the surface. The antennules are specialized with flagella that contain both chemical and mechanical receptors. Information is gathered from the immediate environment and integrated directly into the nervous system. A pair of formidable forward facing rostral horns protects the stalked eyes. To accommodate body growth the entire structure is molted about twice a year. The structure is cast into the environment in much the same way that birds molt their feathers.

The entire carapace, including legs, antennae, mouth parts and eye stalks, is made of chitin. Chitin is one of the most widely distributed fibrous materials in animals and is especially abundant among the arthropods. It is similar in structure to cellulose fibers in plants. Chemically, chitin is a polymer made from repeating units of glucosamine, a nitrogen-containing sugar derivative, linked together by special (glycoside) bonds. The adjacent polymer chains run in different directions. The reason may be that one form (antiparallel) is energetically more favorable than the alternative (parallel) configuration. There are no discernable functional differences among the forms and one form can be converted to the other under harsh chemical treatment resulting in a change of length. The material, which is entirely resynthesized in each molt, is quite hard, but becomes brittle when dried.

An estimated 100 billion tons (10 gigatons!) of chitin is synthesized, molted, and degraded annually. Only small amounts accumulate in the environment. In the ocean bacteria of the genus *Vibrio* reduce the chitin to simple sugars and ammonia. Besides lobsters, crabs, and other crustaceans, chitin occurs in beetles, spiders, worm egg-cases and was first discovered in mushrooms in 1811. It is all recycled through bacteria. Chitosan is a chitin-derivative recovered from the exoskeletons of shrimps and horseshoe crabs, (*Limulus polyphemus*). Easily purified, it is used in medicine to coat sutures as it accelerates healing of burns and surface wounds, reduces pain, and is not rejected by the body. It is used as a filter and to bind metals in water purification. Chitosan, like glucosamine, shows up in human dietary supplements, in cosmetics, toothpaste, and animal feeds. It adds fiber to the diet along with plant cellulose. In some dietary-fiber products chitin is coupled with the lactose-rich, often indigestible milk solid whey to provide bacteria to stimulate digestion.

Across the Gulf Stream, on the eastern Atlantic seaboard, Atlantic horseshoe crabs appear in enormous numbers to breed each spring. Their shells have become a major source of commercial chitin. Horseshoe crabs do not have hemoglobin to carry oxygen in their blood, but instead use hemocyanin, which contains copper, making their blood blue. This relative of spiders and scorpions is also harvested for its proteins, which are used in vision research or as a source of LAL, a protein that is useful as a test for bacterial endotoxins in human blood. However, probably the greatest threat to the horseshoe crabs lies in the lack of nesting sites and the harvest of the adults for bait. Fewer eggs mean less successful reproduction which has reduced the horseshoe crab populations. Plus, the eggs also provide a high-energy food for large numbers of migrating shorebirds. As a consequence, numbers of both red knot (*Calidris canutus*), and ruddy turnstone (*Arenaria interpres*), have fallen dramatically in recent years. Both New Jersey and Delaware have recently passed laws to regulate the harvest of horseshoe crabs and to protect the feeding grounds of these migratory birds.

Gray Triggerfish
(*Balistes capriscus*)

The laterally compressed, ovoid body of the gray triggerfish grows to 30 in. There are two dorsal and two ventral fins, the first having three spines. It can wedge itself into a tight crevice and "trigger" the dorsal fins into an erect position. The mature gray triggerfish is predominantly pale gray, green-gray to yellow-brown with indistinct, irregular and broad dark bars on the body. There is a narrow pale streak on the chin. The mouth is thick-lipped, covering the eight sharp teeth. There are small light blue spots on upper half of body and median fins, and irregular short lines ventrally.

Distribution is primarily shallow waters from Nova Scotia to Bermuda, and south to Argentina. They also occur in the Mediterranean and south to Angola.

Catesby illustrated this fish and called it "*The Old Wife.*" He conjectured on its unique "trigger" defense when being attacked from behind:

> "…tho' Nature seems not to have left them altogether defenceless, their Enemies generally evade the Danger of their Weapons by biting the hind part of the Body short off, but as the Nature of all rapacious Animals is to pursue and devour with furious Eagerness, I conjecture that sometimes by advancing a little too far they are caught by these sharp Bones, one entering the upper, and the other the lower Jaw, which keeps the Mouth from closing, the Consequence of which is that the Devourer will soon be drowned except he can instantly extricate himself from his Prey…" (II-22).

I painted this fish from one I caught on South Water Cay, Belize. It surprised me that, when caught, it produces an "*oink*" sound.

Balistes capriscus · gray trigger fish · 3 dorsal spines — In open waters — fish is gray — can change color
M J Brush   georgia   Average 26 dorsal rays —   to match coral bottom — natural markings

Gray Triggerfish

The peculiar body shape of the gray triggerfish (*Balistes capriscus*) is referred to as laterally compressed (Plate 29, pg. 155). Leatherjacket, an alternative common name, refers to the tough leathery skin. Gray triggerfish prefer reefs and hard-bottom areas. Reefs provide numerous nooks and crannies where it can retreat when threatened. The eye is situated far from the relatively small mouth with its eight fused, specialized teeth. Underlying this geometry is a set of strong adductor muscles that make for a forceful bite. As *Balistes capriscus* is a diurnal predator, this configuration of the eyes and mouth give it a clear view in its search for bottom-dwelling shrimp, crabs, sea urchins, sand dollars, sea stars and bivalve mollusks.

Gray triggerfish move away from their home to more open-bottom areas to feed. Movement is powered and controlled by the dorsal and anal fins rather than the tail fins, as is common in so many other fish. Underway, they resemble a slowly floating zeppelin rather than a streamlined, fast swimming tuna (Plate 25, pg. 139). When feeding they maintain vertical position just inches over the bottom. While hovering they direct a stream of water from their mouth with just enough force to uncover a sand dollar buried just below the surface. If nothing appears, they move a small distance and repeat. When prey is found they increase the force of the water stream until the sand dollar is fully exposed, and then employ the beak-like teeth to flip the sand dollar out onto the surface. They play with it until it is turned over, and then the triggerfish uses a closed mouth to thrust down and crush the exposed center, the most vulnerable point. It is the soft internal organs that provide the meal.

A similar strategy is applied to sea urchins. Clearly the urchin spines provide an adequate defense from frontal attack. The triggerfish, one of the few animals to eat urchins, directs a stream of water underneath the urchin until it tips over. The fish then grabs the unprotected mouthparts and with rapid shakes of its head, shreds the urchin. Triggerfish also feed on small crabs by grabbing the victim directly with a crushing bite on the anterior aspect of the carapace, often biting smaller animals in two. Not surprisingly, this is very effective machinery, based on muscle power, modulated by an efficient nervous system that coordinates the fins to control body position, the production and strength of the water stream, and the actions of the jaw.

Juvenile gray triggerfish are associated with Sargassum and are subject to predation by tuna, dolphinfish (*Coryphaena* spp), marlin (*Makaira* spp), and sharks. Predators, such as amberjack (*Seriola* spp), and larger grouper and sharks, take adults closer to reefs. The first dorsal fin is crucial to its defense as the fish moves slowly and would have a hard time out-swimming a determined predator. The anterior-most spine of the dorsal fin is thick, powerful and almost horn-like. When threatened, the fish will slip into a tight crevice, wedge itself as deeply as possible, and erect the spine. When danger is past, muscles are used to depress the second spine. As the second spine is depressed, the first spine follows passively and again lies flat against the body.

Happily, the gray triggerfish is not considered vulnerable or threatened. It is taken commercially, recreationally for food, and is frequently on display in public aquaria. In some tropical areas, it has been linked to cases of ciguatera poisoning, a condition caused by consumption of fish containing a common dinoflagellate which easily makes its way into the food web.

Sargasso Sea

In the Atlantic Ocean near Bermuda there is an area roughly size of the continental United States called the Sargasso Sea (Plate 30, pg. 159). Called "an ocean within an ocean," its boundaries are determined by a clock-wise flow of currents. The surface is characterized by light winds and little rain. Mariners recognized the area where ships were often becalmed as the "Horse Latitudes". The moniker "Horse Latitudes" derived from the fact that becalmed voyagers would sacrifice their horses for food and to conserve water. Vast amounts of seaweed (marine algae) typically float at the surface. Early Portuguese sailors called the weed "*sargaco*" after the air-filled bladders that resembled the grapes they knew from land. This continuously moving raft of seaweed can be thick, and provide habitat for an amazing diversity of small animals. Unlike other seaweeds these are not attached to the substrate. "*Sargaco*" consists of two species *Sargassum natans* and *S. fluitans*. *S. natans*, also called common or spiny gulfweed, is predominant, with a compact bushy growth pattern, and a single protruding spine on the float. *S. fluitans*, commonly called broad toothed gulfweed, has longer, wider toothy fronds and no spines on the floats.

The Sargasso Sea spans 35º of longitude east to west and about 50º latitude north to south, bounds which can vary as the surface currents that hold it in place move slightly. Clockwise, these are the Florida current to the southwest, the Gulf Stream to the north and northwest, the North Atlantic current to the north and northeast and the Canary Current to the east. The North Equatorial Drift defines the entire southern border. The strength of the currents isolates the area from the rest of the Atlantic Ocean and produces two unique features. One is the rotation of the mass of water, driven by the surrounding currents, which are separated from the deeper seas by a strong temperature gradient. This temperature difference and the rotary motion pile up the water in a lens with its center two ft. higher than the periphery. Second is the temperature structure, which prevents upwelling of nutrients from below. Accordingly, the water remains clear to great depths even though there is some subsurface plankton and algae. Physical isolation caused by the currents prevents much seaweed from coming in or leaving. The seaweed reproduces asexually by vegetative budding. Hence the entire ecosystem is essentially isolated and self-contained.

Sir John Murray (1841-1914), sailing on the *Michael Sars* in 1910, demonstrated that seaweed covered the surface of the Sargasso Sea only in patches. Previously, stories of mariners entrapped in the vast morass of Sargasso weed were prevalent and had been retold since first mentioned by Columbus. Murray had been a naturalist on H. M. S. *Challenger*, and was the main author of the 50-volume *Report of the Scientific Results of the Voyage of the H.M.S. Challenger During the Years 1872-1876*. He was wealthy enough to help subsidize its publication. Subsequent to his observations, it was determined that the patchiness of seaweed was the result of weak winds blowing over long distances interacting with small surface-water currents to produce long avenues of counter-current eddies with bands of sinking water between them. These sinking patterns concentrated the *Sargassum* into long characteristic bands.

Sargasso Sea
(*Sargassum natans*)

Sargassum natans is a marine brown alga, orange to brown with branches bearing flat elongated 1-4 in. narrow leaf-like frond with serrated broad teeth on the margins. It has small air-filled bladders, with a single spine, growing in the axils of fonds. Vast mats of this species are found around Bermuda in the North Atlantic, south to the Caribbean and Central America. This floating world provides food, refuge, and breeding grounds for fish, sea turtles, crabs, shrimps, and more. From the top of the watercolor, left to right, I have painted the following:

Juvenile green Sea Turtle (*Chelonia mydas*)

Atlantic flyingfish (*Cypselurus melanurus*). Catesby painted this fish and wrote:

> "As they are a Prey to both Fish and Fowl, Nature has given them those large Finns, which serve them not only for Swimming, but likewise for Flight. They are good eating Fish, and are caught plentifully on the Coasts of Barbados, where at certain Seasons of the Year the Markets are supplied with them" (II-8).

Portuguese man o' war (*Physalia physalis*)

Fringed filefish (*Monacanthus ciliatus*)

Sargassum pipe fish (*Syngnathus pelagicus*)

Sargasso crab, (*Portunus sayi*)

Planehead filefish (*Stephanolepis hispides*)

Sargassum shrimp (*Leander tenuicornis*)

Sargasso frogfish, (*Histrio histrio*)

Lined seahorse (*Hippocampus erectus*)

Dolphin fish (*Coryphaena hippurus*)

This watercolor is not painted exactly to scale; it is meant as a representation of the abundant life found amongst floating mats of *Sargassum*.

MJ Brush
9/07 Mystic CT. Sargassum natans / Dolphinfish –
Man-o-War –
Frog fish - Histrio histrio
Seahorse - Hippocampus erectus
(plain handed)
Filefish - Stephanolepis hispidus
Flying fish - Parexocoetus brachypterus
Sargasso crab - Portunus sayi
Shrimp — Lsander tenuicornis

The world here is characterized by clear water, bright sunlight and moderate temperatures. Life is abundant in the forest that floats at the surface. The diversity of both resident vertebrates and invertebrates is unique. In addition, many species occur here temporarily as the *Sargassum* forest functions as a nursery, shelter, or breeding ground during appropriate periods for species with more elaborate life cycles. The forest's multifaceted morphology and color pallet of browns and yellows (not greens) provide sanctuary to numerous animals of many different sizes. The *Sargassum* forest provides a floating habitat, or nursery, for fish, crabs, shrimp, bryozoans and nudibranchs. All are relatively small and typically have colors and surface textures that match the background substrate. They move with stealth, patrolling for the copepods, abundant zooplankton, and other larval and juvenile forms that provide their sustenance. Commercially important dolphinfish, jacks, and amberjacks feed here and the Sargasso serves as a nursery for both the green sea and loggerhead turtles.

The endemic fish display an even greater array of exaggerated camouflage. Fins and body parts of the lined seahorse (*Hippocampus erectus*) exquisitely mimic plant structures. The filefish (*Monacanthus ciliatus*), and (*Stephanolapis hispides*), with exotic shapes, appendages, and colors, inhabit the interstices of the seaweed. The Sargasso frogfish (*Histrio histrio*), and others have a compressed body shape with an irregular profile, together with a mottled body surface with light and dark spots and blotches of color. The skin contains pigment-filled chromophores, which like those in the squid and octopus, are under hormonal and nervous control. The color matches that of the *Sargassum* and can be modified depending on light levels and the environment of the fish. The hand-like pectoral fins have evolved into grasping organs used to "climb" through the *Sargassum*.

William Beebe (1877-1962) was a pioneer explorer of the Sargasso Sea. In 1926 he had the steam yacht *Arcturus* retrofitted as a research vessel with laboratories, facilities for trawling and dredging, plus instruments to acquire water samples and measure temperature at various depths. In two visits to the area Beebe recovered specimens that provided the first scientific inventory of the animals associated with Sargassum weed, and also recorded invaluable physical measurements. Beebe, like Murray failed to encounter the dense mats of weed that in legend had entangled ships and dragged unlucky sailors to their doom.

Light In The Darkness Of The Deep

Mark Catesby sailed from England for Carolina early in 1722. One hundred fifty years later in 1872, HMS *Challenger* departed Portsmouth for what was to be the first major voyage of research on the world's oceans (the term oceanography did not yet exist). The expedition lasted almost four years and covered 68,900 miles. The laying of the first trans-Atlantic cables 20 years earlier had piqued interest in the topology of the ocean floor, and there was keen interest in the physical properties of deep ocean water. At that time, dredging was limited to near-shore and essentially no information existed on deep water temperature, density, chemical composition, or light penetration.

Of equal interest was the possibility of sampling life in the depths of the ocean. In the mid-nineteenth century Edward Forbes (1815-1854), hypothesized that no life should exist at great depth (more than 300 fathoms) because of the dire physical conditions and absence of light. Known as the Azoic Hypothesis, it was derived from Forbes' crude trawling results and limited dredging in the relatively shallow Mediterranean Sea. He was attempting to determine if life in the sea was zoned in ways analogous to that on land, as was established by explorer-naturalists such as Alexander von Humboldt and Alfred Russel Wallace. Forbes's dilemma, as he found few animals at depth, was actually the consequence of his inefficient sampling techniques. There was contrary evidence from the experience with trans-oceanic cables. Early on, *Challenger* recovered crinoids (a type of stalked echinoderm) from depths exceeding 6,000 ft., which confirmed life at great depth. Even better, crinoids have an abundant fossil record that reaches back to Jurassic times, the Age of Dinosaurs. This find was particularly important as it also tested one of Darwin's hypotheses that the deep oceans would be home to life forms found on land only as fossils. Darwin's reasoning was that, since the ocean bottom over time was basically unchanged and unchanging, selection would be relaxed and perhaps even non-existent. Hence, crinoids abundant as fossils on land would be unchanged or would resemble their Jurassic forbearers as there would be no selective pressure to change. The presumption was that with extreme environmental stability, evolutionary stasis would certainly follow.

HMS *Challenger's* major tools for undersea exploration were dredges and sounding devices. Dredging could reveal what lived on the bottom and provide samples to reconstruct geological history. Mariners had traditionally used lead lines to gauge depth. By the mid-1800s many technical difficulties with lead lines were resolved. The line was modified so as to stop exactly when the weight hit bottom. The weight was modified to keep the line vertical and eliminate inaccuracies due to the curvature of the line in the water column as the ship moved. Also, the weight could be detached and the line retrieved without breaking. As early as 1854 the US Navy had used this technique to construct a crude undersea geography of the North Atlantic. At first, *Challenger* employed piano-wire sounding lines, which were deployed with each event, hard work even with the help of a motor-driven drum. Quickly, however, the wire mechanism was replaced by more traditional methods. Soundings were taken and dredging secured every 200 miles. Despite the primitive equipment compared with today's multibeam SONAR coupled with satellite GPS, *Challenger* significantly increased understanding of seafloor geology and topography.

Deep East Expedition
September 8 – October 1, 2001

Knowing that Catesby crossed the Atlantic by ship four times in the eighteenth century, I was honored to be included as an artist on the "Deep East" expedition, perhaps as honored as Mr. Catesby was 300 years earlier.

The Deep East Expedition's goal was to explore – using the manned submersible *Alvin* – the floor of the Atlantic, collecting video footage, measuring the biological, geological, and chemical features, and collecting samples for further analysis. During the expedition, scientists examined deep-water corals and methane hydrates, and discovered previously unknown deep-sea resources and processes.

This is a portion of the 4 x 6 ft. painting I created, using sketches made from live specimens collected by *Alvin*, as well as shipboard video, still photographs, and in consultation with the scientists involved. Scientists are always curious.

In September 2016, President Obama designated the Northeast Canyons and Seamounts Marine National Monument, protecting a portion of this area for the future. The specimens collected and the area explored during this expedition directly influenced this proclamation.

Deep Sea Fish

Bristlemouth (*Gonostoma elongatum*) grow to 10 1/2 in. and are found worldwide at depths to three-quarters of a mile in tropical and subtropical waters. It is black with silver markings laterally, and with numerous light-bearing organs called photophores.

The 14 in. threadfin dragonfish (*Echiostoma barbatum*) is found at depths to two miles. It is dark brown in color, has numerous photophores along its sides and sharply pointed teeth. A second single photophore is at the tip of a thin appendage, called a barbell, below the jaw. It may be used to lure prey close to its numerous needle-like teeth. A third photophore is located just back of the eye, larger in the male than the female.

Catesby illustrated a viperfish (*Chauliodus sloani*) in the appendix of the second volume of Natural History although he did not write anything about this fish. The specimen was sent to him by his brother, John Catesby, from Gibraltar.

I illustrated these fish from the collections in the laboratory of Dr. William Kirby-Smith at the Duke Marine Lab, Beaufort, North Carolina. The fish were both collected during a 1966 trawl by Duke Marine Laboratories Research vessel *Eastward*.

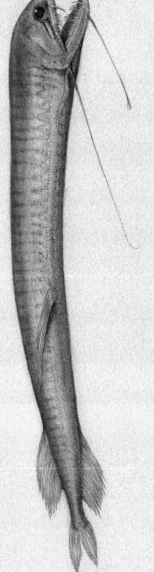

Bristlemouths — *Gonostoma elongatum* — *Echiostoma barbatum*

Undersea exploration grew dramatically with the use of submersibles. In 1934, William Beebe (of the *Arcturus*) descended 3,028 ft. in his tethered bathysphere near Bermuda. The depth tested both the vessel's structural limits and the play and recovery of the tether system. Aside from going where "no man has gone before," Beebe directly observed marine life and compared it with what had been recovered from previous trawling work, including the massive *Challenger* collection. His explorations showed that many of the regions at various depths were more populated, especially with large vertebrates, than the trawling operations had ever yielded. But it was *Alvin*, the self-propelled, titanium-constructed submersible from Woods Hole Oceanographic Institute that eventually succeeded. In a series of explorations, several of which included places where *Challenger* worked, *Alvin* made numerous significant discoveries. In the early 1970s *Alvin* dives detailed the Mid-Atlantic Ridge where geological sampling provided evidence for sea-floor spreading and a mechanism related to the theory of plate tectonics, all of which provided key support for the then emerging idea of continental drift. Investigators also verified the existence of underwater vents called "black smokers". The vents support an active, self-contained, high-temperature ecosystem based on sulfur as the energy source. In addition to extensive work on undersea features such as seamounts, *Alvin* was used to recover an atomic device from the bottom of the Mediterranean and surveyed the wreck of the R. M. S. *Titanic*.

In 2001, M.J. accompanied the National Oceanic and Atmospheric Administration (NOAA) on the Deep East Expedition during the first leg of its field season. Ivar Babb and Dr. Peter Auster at the University of Connecticut's Avery Point campus requested that she create a painting, created from thousands of digital frames taken from the *Alvin* submersible, depicting the deep-sea coral communities in the Georges Bank Canyons. These canyons lie approximately one mile deep in the Atlantic Ocean, 100 nautical miles east of Nantucket (Plate 31, pg. 163).

Deep Sea Fish

Deep-sea organisms live in perpetual darkness. Disproportionately large eyes, large mouths, and strange body shapes, presumably adaptations to very low light levels and limited energy availability, characterize many deep-sea fish. Both the elongated bristlemouth (*Gonostoma elongatum*) and the threadfin dragonfish (*Echiostoma barbatum*) (Plate 32, pg. 165), are included in the family Stomiidae along with other bristlemouths, lightfish, hatchet fish, snaggletooth, viperfish, black dragonfish (or sawtail fish), loose jaws, and scaly and scaleless dragonfish. All these descriptively named fish are bathypelagic forms, inhabitants of those deepest parts of the ocean beyond the reach of sunlight. There are multiple body appendages, some of which bear light organs, as well as patterned rows of light organs. These photophores are used as lures to attract prey and for signaling in the dark. Signals are used in species recognition and mating displays. The task of finding mates in the dark is obviously difficult and many species are hermaphroditic.

Bioluminescence is primarily a marine phenomenon and is the largest source of light in the deep ocean. "Living light" is generated by a broad variety of animals. Fish, shrimp, jellyfish, and even the unicellular marine dinoflagellates are capable of producing light. A chemical reaction in the photophores is responsible. Just as in the terrestrial firefly, light production depends on a light-emitting pigment (luciferin) activated by an enzyme (luciferase). The emitted light is in the blue range (440-470nm), which matches the optical transparency of ocean water. The pattern of flashing can encode information that serves to warn or to attract mates or prey. Many species of squid, jellyfish, and siphonophores, all emitters, are transparent and colonial, which serves to intensify the signal.

Homeward Bound

When Catesby left the Bahamas and sailed to England, his experience as a field naturalist ended, and the second half of his life as an artist, author, and businessman began. His legacy, created over the next 23 years, is his two magnificent volumes. We trust you have enjoyed our journey in his footsteps.

Mark Catesby's death notice, written by Peter Collinson, appeared in *The Gentleman's Magazine* in 1750:

> "On Saturday morning the 23d of December, died at his house behind St Luke's church, in Old street, the truly honest, ingenious, and modest Mr Mark Catesby, F.R.S. who, after travelling through many of the British dominions, on the continent, and in the islands of America, in order to make himself acquainted with the customs and manners of the natives, and to collect observations on the animals and vegetables of those countries which he there very exactly delineated, and painted on the spot, he returned with these curious materials to England, and compiled a most magnificent work, intitled, A Natural History of Carolina, Florida and the Bahama Islands, which does great honor to his native country, and perhaps is the most elegant performance of its kind that has yet been published since not only the rare beasts, birds, fishes, and plants &c were drawn, engraven, and exquisitely coloured from his original paint by his own hands, in 220 folio copperplates; but he has also added a correct map, and a general natural history of that world. He liv'd to the age of 70, well known to, and much esteemed by, the curious of this and other nations, and died much lamented by his friends, leaving behind him two children and a widow, who has a few copies of this noble work, undisposed of."

ACKNOWLEDGEMENTS

We are grateful to the colleagues, reviewers, and editors who read all or parts of this manuscript during its generation. These include Peter Auster (University of Connecticut), Robert Askins (Connecticut College), Helen Rodwansky (University of Connecticut), the late Peter Stettenhiem, and others who have helped. We also received comments and advice from Lisa D. Brush (University of Pittsburgh), who helped formulate ideas, focused our thinking, and contributed in innumerable other ways. Ed Hagenstein, Jane O'Donnell, and Virginia Bitting provided editorial assistance. We thank them all. We appreciate Jean Thompson Black for her interest and encouragement particularly in the early years. The Biodiversity Heritage Library (www.biodiversitylibrary.org) provided digital access to their first edition of Catesby.

David Elliott of the Catesby Commemorative Trust provided strong support and encouragement. His recommendations, guidance, and administrative skill have been elemental in enhancing the Catesby legacy.

The editorial skills of Charles Nelson of Enniskellen, Northern Ireland, improved the accuracy and clarity of our work. His comprehensive knowledge of the primary Catesby material is unparalleled.

There are countless harbormasters and dockhands from Maine to Florida who helped during our voyages aboard *Mokita* our Cape Dory 330. Emeritus status at the University of Connecticut provided unlimited access to the University Library and its databases. The Dodd Center Library at the University provided access to their Catesby volumes. The curators at the Department of Ornithology at the American Museum of Natural History allowed M.J. unrestricted access to their collections. We thank the libraries, collections and curators at Connecticut College and Yale University Peabody Museum for access to various specimens.

At Duke Marine Lab in Beaufort, NC, W. Kirby-Smith graciously provided laboratory facilities. The administrative staff saw to our accommodations, and faculty provided intellectual spice and the entire group engaged in some serious conversations on women's basketball.

A special thanks to the Guild of Natural Science Illustrators (gnsi.org), for promoting the skills needed in scientific illustration. This organization has given all of us, who seek to illuminate our natural world, a wonderful environment for continuing education and long friendships.

Lastly, we thank all the scientists, explorers and artists that preceded and will follow us. We honor the spirit of Mark Catesby, a truly remarkable man.

Appendix
Resources and Further Reading

Following is a list of books and articles to provide enhanced access to background material and additional reading for the curious.

Natural History Past and Present

Birkhead, T. 2008. *The Wisdom of Birds. An illustrated History of Ornithology.* 413 pgs. Bloomsbury, New York, NY.

Fishman, G. 2000. *Journeys through Paradise. Pioneering Naturalists in the Southeast.* 306 pgs. University of Florida Press, Gainesville, FL.

Frick, G. & R.P. Stearns. 1961. *Mark Catesby, The Colonial Audubon.* 137 pgs. Universary of Illinois Press, Urbana, IL.

Huxley, R. (Ed) 2007. *The Great Naturalists.* 304 pgs. The Natural History Museum. London, UK.

Kastner, J. 1977. *A Species of Eternity.* xiv + 359 pgs. A.A. Knopf, New York

Larson, A. L. 1993. *Not Since Noah: The English Scientific Zoologists and the Craft of Collecting,* 1800-1840. 403 pgs. Unpublished PhD Dissertation. Princeton University Press.

Rice, T. 1999 *Voyages of Discovery. Three Centuries of Natural History Exploration.* 335 pgs. The Natural History Museum, Clarkson Potter/Publishers, London UK.

Riley, M. 2006. The Club at the Temple Coffee House Revisited. *Archives of Natural History* 33(1): 90-100.

Schiebinger, L. 2004. *Plants and Empire: Colonial Bioprospecting in the Atlantic World.* x+306 pgs. Harvard University Press, Cambridge, MA.

Thomson, K. 2005. *Before Darwin: Reconciling God and Nature.* xiv+314 pgs. Yale University Press, New Haven, CT.

Young, D.1992. *The Discovery of Evolution.* Natural History Museum Publications, Cambridge University. Press, London, UK.

And Selected Others On Mark Catesby

Aitken, R. 2007. *Botanical Riches.* xii+243 pgs. Lund Humphries, Burlington, VT.

Allen, E.G. 1937. New Light on Mark Catesby. *Auk* 54:349-363

Allen, E.G.1951. The History of American Ornithology before Audubon. *Trans. Amer. Phil. Soc.* 41(3):387-591. Reprinted (1977) by W. Graham Arader III, King of Prussia, PA.

Cafferty, S. 2007. *Mark Catesby. Colonial Naturalist and Artist*. pgs. 124-132 in Huxley, R., Ed. The Great Naturalists. 304 pgs. Thames & Hudson, New York, NY.

Feduccia, A.1985. *Catesby's Birds of Colonial America*. xvi+176 pgs. University of. North Carolina Press, Chapel Hill, NC.

McBurney, H. 1997. *Mark Catesby's Natural History of America. The Watercolors from the Royal Library, Windsor Castle*. 160 pg. Merrell Holberton Pubs., London, UK.

Meyers, A.R.W. & M.B. Pritchard. 1998. *Empire's Nature. Mark Catesby's New World Vision*. xviii+272. University of North Carolina Press, Chapel Hill, NC.

Neal, C. & D.J. Elliott (Directors) 2007. *The Curious Mark Catesby*. DVD, 57 minutes. Marsh Elder Productions.

Nelson, E.C. & D.J. Elliott. 2015. *The Curious Mister Catesby*. xviii+425 pgs. University of Georgia Press. Athens, GA.

Stone, W. 1929. Mark Catesby and the Nomenclature of North American Birds. *Auk* XLVI: 447-454

Wilson, D. 1971. The Iconography of Mark Catesby. *Eighteenth-Century Studies*. 4(2): 169-183.

Collecting and Classifying: Organizing The Natural World

Blunt, W. 2001. *Linnaeus. The Compleat Naturalist*. 264 pg. Princeton University Press, Princeton, NJ.

Brooks, D. R. & D. A. McLennan. 1991. *Phylogeny, Ecology, and Behavior. A Research Program in Comparative Biology*. xii+ 434pg. University of Chicago Press. Chicago, IL.

Daston, L. 2004. Type Specimens and Scientific Memory. *Critical Inquiry* 35: 153-182

Endersby, J. 2008. *Imperial Nature. Joseph Hooker and the Practice of Victorian Science*. xii+429 pgs. University of Chicago Press, Chicago, IL.

Fara, P. 2003a. Carl Linnaeus: Pictures and Propaganda. *Endeavour* 27(1): 14-15.

Fara, P. 2003b. *Sex, Botany and Empire*. 168 pg. Columbia University Press, New York, NY.

Mayr, E. 1969. *Principles of Systematic Zoology*. x+428 pgs. McGraw-Hill, New York, NY.

Mayr, E. 1982. *The Growth of Biological Thought. Diversity, Evolution and Inheritance*. 974 pgs. Belknap Press of Harvard University Press. Cambridge, MA.

Moser, Barry 1999. *The Holy Bible: King James Version, Pennyroyal Caxton Bible*, 993 pgs. + List of illustrations. Viking Studio Edition, Penguin Publishing, Putnam, N.Y.

O'Brian, P. 1997. *Joseph Banks: A life*. 326 pgs. University of Chicago Press, Chicago, IL.

Padian, K. 1999. Charles Darwin's Views of Classification in Theory and Practice. *Syst. Biol.* 48(2): 352-364.

Reveal, J.L. 2012. A Nomenclatural Summary of the Plant and Animal Names Based on Images in Mark Catesby's Natural History (1729-747). *Phytoneuron* 2012-11:1-32.

Schuh, R. T. 2003. The Linnaean System and Its 250-year Persistence. *Botanical Review* 69(1): 59-78.

Stearn, W. T. 1958. Botanical Exploration to the Time of Linnaeus. *Proc. Linn. Soc. London.* 169:173-19

Stone, W. 1929. Mark Catesby and the Nomenclature of North American Birds. *Auk* XLVI: 447-454.

Thomson, K. 2005. *Before Darwin. Reconciling God and Nature.* xiv+314 pg. Yale University Press, New Haven, CT.

Walters, M. 2003. *A Concise History of Ornithology.* Yale University Press. New Haven, CT.

Magnificent Magnolias

Aitken, R. 2007. *Botanical Riches: stories of botanical exploration.* xi+243 pgs. Aldershot: Lund Humphries/Ashgate. Burlington, VT.

Nelson, E.C. 2014. Georg Dionsius Ehret, Mark Catesby and Sir Charles Wagner's Magnolia grandiflora: an early eighteenth-century picture puzzle resolved. *Rhododendrons, camellias and magnolias.* 65: 36-51

Ivory-billed Woodpecker

Askins, R.A. 2000. *Restoring North America's Birds.* xiii+320 pgs. Yale University. Press, New Haven, CT.

Hoose, P. 2004. *The Race to Save the Lord God Bird.* 208 pgs. Farrer, Straus & Giroux

Jackson, J.A. 2002. *Ivory-billed Woodpecker* (*Campephilus principalis*). *in* The Birds of North America, No 711 (A. Poole and F. Gill, Eds.). The Academy of Natural Science, Philadelphia, PA. and The American Ornithologists' Union, Washington, D.C.

Jackson, J. A. 2004. *In search of the Ivory-billed Woodpecker.* x+294 pg. Smithsonian Press, Washington, DC.

Reverend Bachman's warbler

Barrow, M.V. 1998. *A Passion for Birds. American Ornithology after Audubon.* x+326 pgs. Princeton University Press. Princeton, NJ.

Bartram, W. 1791. *Travels through North & South Carolina, Georgia, East & West Florida.* From the 1955 Dover edition which is a reprint of the Macy-Masius 1928 edition edited by M. Van Doren.

Brown, R.E. & J.G. Dickson. 1994. *Swainson's Warbler (Limnothlypis swainsonii). in* The Birds of North America, No 126 (A. Poole and F. Gill, Eds.). The Academy of Natural Sciences, Philadelphia, PA. and the American Ornithologists' Union, Washington, DC.

Burtt, E.H. & W.E. Davis. 2013. *Alexander Wilson. The Scot Who Founded American Ornithology.* ix+444 pgs. Belknap Press of Harvard University Press, Cambridge, MA.

Hamel, P. B. 1995. *Bachman's Warbler (Vermivora bachmanii). in* The Birds of North America, No 150 (A Poole and F Gill, Eds.). The Academy of Natural Sciences, Philadelphia, PA and the American Ornithologists' Union, Washington, DC.

Mearns, B. & R. Mearns.1992. *Audubon to Xantus. The Lives of Those Commemorated in North American Bird Names.* 588 pgs. Academic Press. San Diego, CA.

Schuler, J. 1995. *Had I the Wings. The Friendship of Bachman & Audubon.* xii+233 pgs. University of Georgia Press. Athens, GA.

Passenger Pigeon

Cokinos, C. 2000. *Hope is the Thing with Feathers.* 359 pgs. Jeremy P. Tarcher/Putnam. NY.

Feduccia, A. 1985. *Catesby's Birds of Colonial America.* xvi+176 pg. University of North Carolina Press, Chapel Hill, NC.

Pimm, S., P. Raven, A. Peterson, C. H. Sekercioglu, & P. R. Ehrlich. 2006. Human impacts on the rates of recent, present, and future bird extinctions. *Proc. Nat. Acad. Sci.* 103: 10941-10946.

Schorger, A. W. 1952. Introduction of the Domestic Pigeon. *Auk* 69: 462-463

Terborgh, J. 1989. *Where Have All the Birds Gone?* xvi+207 pgs. Princeton University Press. Princeton, NJ.

Rice

Carney, J. 2005. Out of Africa. Colonial Rice History in the Black Atlantic. pgs 204-207 *in Colonial Botany. Science, Commerce, and Politics in the Early Modern World.* L. Schiebinger and C. Swan, Eds. University of Pennsylvania Press, Philadelphia, PA.

Edgar, W. 1998. *South Carolina, a History.* 716 pgs. University of South Carolina Press, Columbia, SC.

Rosset, R., J. Collins, and F. M. Lappe. 2000. Lessons from the Green Revolution. *Tikkim Magazine.* Institute for Food and Development Policy.

http://www.foodfirst.org

Seeing the Forest and the Trees

Anderson, S.H., D. Kelly, J. J. Ladley, S. Molloy, J. Terry 2011 Cascading Effects of Bird Functional Extinction Reduce Pollination and Plant Diversity. *Science* 331: 1068-1071.

Davis, M.A. 2009. *Invasion Biology.* Oxford University. Press. ix+288 pgs. Oxford, UK.

Dawkins, R. 2003. *A Devil's Chaplain. Reflections of Hope, Lies, Science and Love.* vi+223 pgs. Houghton Mifflin Co., Boston, Ma.

Drury, W. H. 1998. *Chance and Change. Ecology for Conservationists.* xxiii+223. University of California Press, Berkeley, CA.

Ehrlich, P. & A. Ehrlich 2004. *One with Nineveh. Politics, Consumption, and the Human Future.* 447 pgs. Island Press, Washington, DC.

Foster, D. R. & J. D. Aber. 2004. *Forests in Time.* xiv+477 pg. Yale University Press. New Haven, CT.

Hubbell, S.P. 2001. *The Unified Neutral Theory of Biodiversity and Biogeography.* Monographs in Population Biology #32. xiv+375 pgs. Princeton University Press, Princeton, NJ.

Magee, J. 2007. *The Art and Science of William Bartram.* xi+264 pgs. Natural History Museum, London & Pennsylvania State University Press.

McGraw, J. B. & M. A. Furedi. 2005. Deer Browsing and Population Viability of a Forest Understory Plant. *Science* 307:920-922.

Pimm, S. L. 2001. *The World According to Pimm.* A scientist audits the earth. xiii+285 pg. McGraw-Hill, New York, NY.

Russell, E. W. B. 1997. *People and the Land through Time. Linking Ecology and History.* xx+ 306 pgs. Yale University Press, New Haven, CT.

Wennersten, J. R. 1996. Soil Miners Redux: The Chesapeake Environment 1680-1810. *Maryland Historical Magazine* 91(2): 157-179.

Wilcove, D. S. 1999. *The Condor's Shadow.* xix+339 pgs. Anchor Books, New York, NY.

Williams, M. 2003 *Deforesting the Earth. From Prehistory to Global Crisis.* 750 pg. University of Chicago Press, Chicago, IL.

Catalpa

Laird, M. 1998. From Callicarpa to Catalpa: The Impact of Mark Catesby's Plant Introductions on English Gardens of the Eighteenth Century. Pg. 184-227. *in Empire's Nature. Mark Catesby's New World Vision.* A. R. W. Meyers & M. B. Prichard, Eds. University of North Carolina Press xviii+271pp.

Carolina Jessamine

Native Plants Database (http://www. Wildflower.org/about.php)

USDA, NRCS 2015. The PLANTS Database (http://plants.usda.gov) National Plant Data Team, Greensboro, N.C.

An Ecosystem Dependent on Fire

Askins, R. A. 2000 *Restoring North America's Birds*. xiii+320 pgs. Yale University Press. New Haven, CT..

Conner, R. N. & B. A. Locke. 1982. Fungi and Red-cockaded Woodpecker Cavity Trees. *Wilson Bull.* 94:64-70

Jackson, J. A. 1994. *Red-cockaded Woodpecker (Picoides borealis). in* The Birds of North America, No. 85 (A. Poole and F. Gill, Eds.). The Academy of Natural Sciences, Philadelphia. The American Ornithologists' Union, Washington, D.C.:

James, F. C. 1991. Signs of trouble in the largest remaining population of Red-cockaded Woodpeckers. *Auk* 108:419-423

Winkler, H., D. A. Christie, & D. Nurney 1995. *Woodpeckers. An Identification Guide to the Woodpeckers of the World.* 406 pg., 64 colored plates. Houghton Mifflin, NY.

Tulip Tree

American Ornithologists' Union. 1998. *Check-list of North American Birds.* liv+829 pgs. American Ornithologists' Union. Washington, DC.

Ennos, R. 2001. *Trees.* 112 pgs. Smithsonian Institution Press, Washington, DC.

Trudge, C. 2006. *The Tree.* xix+459 pgs. Crown Publishers (a Division of Random House), New York, NY.

Oaks

Mabberley, D. J. 2008. *Mabberley's Plant Book: A Portable Dictionary of Plants, Their Classification, and Uses.* 3rd Edition' 1040 pgs. Cambridge University Press, Cambridge, UK.

Stuart, D. 2002. *The Plants That Shaped our Gardens.* 208 pgs. Harvard University Press. Cambridge, MA.

Thugs and Aliens

Baskin, Y. 2002. *A Plague of Rats and Rubbervines. The Growing Threat of Species Invasion.* vii+371 pg. Island Press/Shearwater Books. Washington, DC.

Coates, P. 2006. *American Perceptions of Immigrant and Invasive Species.* x+256 pg. University of California Press. Berkeley, CA.

Cox, G. W. 2004. *Alien Species and Evolution.* xi+377 pgs. Island Press, Washington, DC.

Flannery, T. 2001. *The Eternal Frontier. An ecological history of North America and its peoples.* 404 pg. Atlantic Monthly Press, New York, NY.

Russell, E. W. B. 1977. *People and the Land through Time: Linking Ecology and History.* xx + 306 pgs. Yale University Press, New Haven, CT.

Simberloff, D. 2013. *Invasive Species. What Everyone Needs to Know.* vii+329pgs. Oxford University Press, New York, NY.

Strauss, S. Y., J. A. Lau & S. P. Carroll. 2006. Evolutionary Responses of natives to introduced species: what do introductions tell us about natural communities? *Ecology Letters* 9: 357-376.

Wilcove, DS 1999. *The Condor's Shadow. The Loss and Recovery of Wildlife in America.* xix+339 pg. Anchor Books, New York, NY.

Brazilian Pepper-Tree

Ferriter, A. (Editor). 1997. *Management Plan for Florida.* The Florida Exotic Pest Plant Brazilian Pepper Task Force. 31 pgs.

Johnston, D. W. 2003. *The History of Ornithology in Virginia.* x+219 pgs. University of Virginia Press. Charlottesville, VA.

Silvertown, J. 2005. *Demons in Eden: The Paradox of Plant Diversity.* x+ 169 pgs. University of Chicago Press. Chicago, IL.

Stuart, D. 2002. *The Plants that shaped our Gardens.* 208 pgs. Harvard University Press. Cambridge, MA.

Honeysuckle/Cedar Waxwing

Hudon, J. & A. H. Brush. 1989. Dietary basis of a color variant of the Cedar Waxwing. *J. Field Ornith.* 60 (3):361-368.

Brush, A. H. 1990. A possible source for the Rhodoxanthin in some Cedar Waxwings. *J. Field Ornith.* 61 (3):355.

Witmer, M. C. 1996 Consequences of an Alien Shrub on the Plumage Coloration and Ecology of Cedar Waxwings. *Auk* 113 (4):735-743.

Purple Loosestrife

Cox, G. W. 2004. *Alien Species and Evolution.* xi+377 pgs. Island Press, Washington, DC.

Russell, E. W. B. 1997. *People and the Land through Time.* xx+306 pg. Yale University Press. New Haven, CT.

Silvertown, J. 2005 *Demons in Eden: The Paradox of Plant Diversity.* x+169 pgs. University of Chicago Press. Chicago, IL.

Salt Marsh: A Dynamic Equilibrium

Bertness, M. D. 1999. *The Ecology of Atlantic Shorelines.* xii+417 pgs. Sinauer Associates, Sunderland, MA.

Reiger, G. 1991. *Wanderer on My Native Shore.* 268 pg. Lyons and Burford, New York, NY.

Silliman, B. R., J. V. Koppel, M. D. Bertness, L. E. Stanton, & I. A. Mendelssohn. 2005. Drought, Snails, and Large-Scale Die-Off of Southern U.S. Salt Marshes. *Science* 310:1803-1806.

Vaernberg, F. J. & W.B. Vaernberg, 2001. *The Coastal Zone. Past, Present, and Future.* xiv+191 pgs. University of South Carolina Press, Columbia, SC.

Ibis: Red & White

Feduccia, A. 1985. *Catesby's Birds of Colonial America.* Color plates + 176 pgs. University of North Carolina Press, Chapel Hill, NC.

Fox, D. L. 1976. *Animal Biochromes and Structural Colours.* xvi+433 pgs. University of California Press. Berkeley, CA.

Hancock, J. A., J. A. Kushlan & M. P. Kahl. 1992. *Storks, Ibises and Spoonbills of the World.* 384 pgs. Academic Press. London, UK.

Hill, G. E. & K. J. McGraw, Eds. 2006. *Bird Coloration.* Vol 1. *Mechanisms and Measurements*, ix+589 pgs. Vol 2 *Function and Evolution*, x+477 pgs. Harvard University Press, Cambridge, MA.

Kushlan, J. A. & K. L. Bildstein. 1992. *White ibis* (*Eudocimus albus*). *in* The Birds of North America, No. 9. (A. Poole, P. Stettenhiem, and F Gill, Eds.) Academy of Natural Sciences, Philadelphia. The American Ornithologists Union, Washington, DC.

Palmer, R. S. (Ed) 1978. *Handbook of North American Birds.* Vol. 1 (Loons through Flamingos). 567 pgs. Yale University Press, New Haven, CT.

Great Blue Heron

Butler, R. W. 1992. *Great Blue Heron* (*Ardea herodias*). *in* The Birds of North America, No. 25 (A Poole, P. Stettenhiem, and F Gill, Eds.). The Academy of Natural Sciences, Philadelphia. The American Ornithologists' Union, Washington, DC.

Hancock, J. A. & J. A. Kushlan. *The Heron Handbook*. 1984. 288 pg. Harper & Row. New York, NY.

Kushlan, J. A. & H. Hefner. 2000 *Heron Conservation*. xvi+480 pg. Academic Press, San Diego, CA.

Marchant, J. H., S. N. Freeman, H. Q. P. Crick & L. P. Beaven. 2004. The BTO Heronries Census of England and Wales 1928-2000: new indices and a comparison of analytical methods. *Ibis* 146:323-334.

Green Heron

Crick, H. Q. P. 2004 The Impact of Climate Change on Birds. *Ibis* 146: 48-56 (Suppl.1).

Davis, W. E., Jr., & J. A. Kushlan.1994 *Green Heron (Butorides virescens). in* The Birds of North America, No. 129 (A. Poole & F Gill, Eds.) The Academy of Natural Science. Philadelphia. The American Ornithologists Union. Washington, DC.

Lovejoy, T. E. & L. Hannah (Eds.). 2005. *Climate Change and Biodiversity*. xiii+418 pgs. Yale University Press. New Haven, CT.

Root, T. L, J. T. Price, K. R. Hall, S. H. Schneider, C. Rosenwein & A. J. Pounds. 2003. Fingerprints of Global Warming on Wild Animals and Plants. *Nature* 421: 57-60

Fish Crow

Audubon, J. J. 1816-1838. *Birds of America* (in 4 volumes).

Banko, PC, D. L. Ball, & W. E. Banko. 2002. *Hawaiian Crow (Corvus hawaiiensis). in* The Birds of North America, No. 648 (A. Poole and F. Gill, eds.). The Academy of Natural Sciences, Philadelphia. The American Ornithologists' Union, Washington, DC.

Kastner, J. 1977. *A Species of Eternity*. xiv+350 pgs. A Knopf Publishers, New York, NY.

LaDeau, S. L., A. M. Kilpatrick & P. P. Marra. 2007. West Nile Virus Emergence and Large-scale Declines of North American Bird Populations. *Nature* 447:710-714.

Madge, S. & H. Brown. 1994. *Crows and Jays: a guide to the crows, jays and magpies of the world.* xxiii+189 pgs. 30 Plates. Houghton Mifflin, Boston, MA.

Marzluff, J. M. & T. Angell. 2005. *In the Company of Crows and Ravens*. xix+384 pgs. Yale University Press, New Haven, CT.

McGowan, K. J. 2001. *Fish Crow (Corvus ossifragus)*. In The Birds of North America, No. 589 (A. Poole and F. Gill, Eds.). The Academy of Natural Sciences, Philadelphia. The American Ornithologists' Union, Washington, DC.

Nuttall, T. 1891. *A Poplar Handbook of the Ornithology of Eastern North America*. Revised and annotated by M. Chamberlain. Originally in 2 volumes Little Brown, Boston, MA.

Wood, M. J. & C. L. Cosgrove. 2006. The hitchhiker's guide to avian malaria. *Trends in Ecology and Evolution.* 21:5-7.

www.cdc.gov/nicidod/dubid/westnile/

American Oystercatcher

McNamara, K. R. 1990. The Feathered Scribe: The Discourse of American Ornithology before 1800. *William & Mary Quarterly.* 3rd Ser. 47 (2): 210-234.

Nol, E. & R. C. Humphrey. 1994. *American Oystercatcher (Haematopus palliatus). in* The Birds of North America, No 82 (A. Poole & F. Gill, Eds). The Academy of Natural Sciences, Philadelphia. The American Ornithologists' Union, Washington, DC.

Zeranski, J. D. & T. R. Baptist.1990 *Connecticut Birds.* xxiii+328 pg. University of New England Press, Hanover, NH.

Find a spot; Settle Down

Chesapeakebay.noaa.gov.oysters/oyster-restoration

Committee on Non-native Oysters in the Chesapeake Bay. (J. Anderson & D. Hedgecock Co-Chairs) 2004. *Non-native Oysters in the Chesapeake Bay.* 344 pg. National Academy Press, Washington, DC ISBN 0-309-09052-0

Feely, R. A., C. L. Sabine, K. Lee, W. Berelson, J. Kleypas, V. J. Fabry, & F. J. Millero. 2004. Impact of Anthropogenic CO_2 on the $CaCO_3$ System in the Oceans. *Science* 305:362-366.

Gutiérrez, J., C. G. Jones, D. L. Strayer & O.O. Iribarne. 2003. Mollusks as ecosystem engineers: the role of shell production in aquatic habitats. *Oikos* 101:79-90.

International Panel on Climate Change. 2007. *Climate Change 2007: The Physical Basis.* Report of the IPCC, 21 pg. Available at http://www.ipcc.ch.

Kennedy. U. S., R. I. E. Newell, & A. F. Eble.1996. *The Eastern Oyster Crassostrea virginica.* Xv+734 pg. Maryland Sea Grant College, College Park, MD.

Kozyr, A., T. Ono, & A. F. Ross. 2004. The Oceanic Sink for Anthropogenic CO_2. *Science* 305:367-376.

Lee, H, N. F. Scherer, & P. B. Messersmith. 2006. Single-molecule Mechanisms of Mussel Adhesion. *Proc. Nat. Acad. Sci.* 103 (39): 12999-13003.

Lenihan, C.H. Peterson, J. E. Byer, J. H. Grabowski, G. W. Thayer & D. R. C. Colby. 2001. Cascading of habitat degeneration: Oyster reefs invaded by refugee fishes escaping stress. *Ecol. App.* 11(3): 764-783

Sabine, C. L., R. A. Feely, N. Gruber, A. M. Key, K. Lee, J. L. Bullister, R. Wanninkhof, C. S. Wong, D. W. R. Wallace, B. Tilbrook, F. J. Millero, T-H Peng, A. Kozyr, T. Ono, & A. F. Ross.

2004. The Oceanic Sink for Anthropogenic CO_2. *Science* 305:367-376.

Oceans formerly full of fish

Berrill, M. 1997. *The Plundered Seas.* x+208 pgs. Sierra Club Books, San Francisco, CA.

Crist, D.T., G. Scowcroft, & J.M. Harding. 2009. *World Ocean Census. A Global Survey of Marine Life.* 256 pgs. Firefly Books

Fahrenkamp-Uppenbrink, J., D. Malakoff, J. Smith, C. Ash, & S. Vignieri. 2015 Oceans of Change. *Science* 350: 750-782.

Froese, R. and D. Pauly, Editors. 2015. WWW.Fishbase.org

Gimlette, J. 2005. *Theatre of Fish. Travels Through Newfoundland and Labrador.* xxii+360 pgs. A.A. Knopf, New York, NY.

Jackson, J. B. C., et al. 2001. Historical Overfishing and the Recent Collapse of Coastal Ecosystems. *Science* 293:629-638.

Jennings, S, M. J. Kaiser & J. D. Reynolds. 2001. *Marine Fisheries Ecology.* xiii+ 417 pgs. Blackwell Sciences, Ltd., Oxford, UK.

Jordan, D.S. 1884. An Identification of the Figures of Fishes in Catesby's Natural history of Carolina, Florida, and the Bahama Islands. *Proc. U.S. National Museum* 7:190-199.

Kurlansky, M. 1997. *Cod. A Biography of the Fish that Changed the World.* 294 pg. Walker and Co., NY.

Kurlansky, M. 2008. *The Last Fish Tale: The Fate of the Atlantic and Survival in Gloucester, America's Oldest Fishing Port and Most Original Town.* 304 pg. Random House/Jonathan Cape. New York, NY.

Murdy, E. O., R. S. Birdsong, & J. A. Musick 1997. *Fishes of the Chesapeake Bay.* xi+324 pg. Smithsonian Institution. Press, Washington, DC.

Myers, R. A., J. K. Baum, T. D. Shepard, S. P. Powers, & C. H. Peterson. 2007. Cascading Effects of the Loss of Apex Predatory Sharks from a Coastal Ocean. *Science* 315: 1846-1850.

Norse, E. A. & L. B. Crowder, Eds. 2005. *Marine Conservation Biology: The Science of Maintaining the Sea's Biodiversity.* 470 pgs. Island Press. Washington, DC.

Pauly, D. & Maclean, J. 2003. *In a Perfect Ocean. The state of fisheries and ecosystems in the North Atlantic Ocean.* xx+175 pg. Island Press, Washington, DC.

Penn, D. J. 2003. The Evolutionary Roots of our Environmental Problems: Towards a Darwinian Ecology. *Quarterly Review of Biology* 78 (3): 275-301.

Peters, D. S. & W. E. Schaaf. 1991. Empirical model of the Trophic Basis for Fishery Yield in Coastal Waters of the Eastern USA. *Transactions American Fisheries Society* 120:459-473

Pimm, S. L. 2004. *The World According to Pimm. A Scientist Audits the Earth.* xiii+285 pg. McGraw-Hill Publishers, New York, NY.

Richardson, A. J. & D. S. Schoeman. 2004. Climate Impact on Plankton Ecosystems in the Northeast Atlantic. *Science* 305: 1609-1612.

Rivinus, E. F., & E. M. Youssef. 1992. *Spencer Baird of the Smithsonian.* X+228 pgs. Smithsonian Institution Press, Washington, DC.

Roberts, C. 2007. *The Unnatural History of the Sea.* xvii+435 pgs. A Shearwater Book, Island Press. Washington, DC.

Vaughn, D. S., M. R. Collins & D. J. Schmidt.1995. Population Characteristics of the Black Sea Bass *Centropristis straita* from the southeastern U.S. *Bulletin of Marine Science.* 56(1):250-267.

Wolf, D. A. 2000. *A History of the Federal Biological Laboratory at Beaufort, NC.* 1899-1999. vii+312. U.S. Dept. Commerce, Washington, DC.

Worm, B, E. B. Barbier, N. Beaumont, J. E. Duffy, C. Folke, B. S. Halpren, J. B. C. Jackson, H. K. Lotze, F. Micheli, S. R. Palumbi, E. Sala, K. A. Selkoe, J. J. Stachowicz, & R. Watson. Impacts of Biodiversity Loss on Ocean Ecosystems Services. *Science* 314: 787-790

Zimmer, C. 2003 Rapid Evolution Can Foil Even the Best-Laid Plans. *Science* 300:895

700 islands

Buden, D. W. 1987. *The Birds of the Southern Bahamas.* BOU Checklist #8. 119 pg. British Ornithologists' Union. London, UK.

Larsen, C. L, et al. 2002. *A Biohistory of Health and Behavior in the Georgia Bight: The Agricultural Transition and the Impact of Contact. in* The Backbone of History: Health and Nutrition in the Western Hemisphere, edited by Richard H. Steckel and Jerome C. Rose. 406-439 pgs. Cambridge University Press, New York, NY., pgs. 406-439.

Mann, C. C. 2002. *1491. Atlantic Monthly.* March, 41-53.

Raffaele, H., J. Wiley, O. Garrido, A. Keith, & J. Raffaele. 1998. *A Guide to the Birds of the West Indies.* 510 pg. Princeton University Press. Princeton, NJ.

Ricklefs, R. E. & E. Birmingham. 2001. Nonequilibrium Diversity Dynamics of the Lesser Antillean Avifauna. *Science* 294:1522-1524.

Schoener, T. W., D. Spiller & J. B. Losos. 2000. Natural Restoration of the Species-area Relation for a Lizard after a Hurricane. *Science* 294:1525-1528

White, A. W. 1998. *A Birder's Guide to the Bahama Islands (Including Turks and Caicos).* x+302 pg. American Birding Association, Colorado Springs, CO.

Sargassum

Earle, S. A. 2001. *National Geographic Atlas of the Ocean.* 192 pgs. National Geographic, Washington, DC.

Gould, C. G. 2004. T*he Remarkable Life of William Beebe. Explorer and Naturalist.* xv+447 pgs. A Shearwater Book. Island Press, Washington, DC.

Matsen, B. 2005. *Descent. The Heroic Discovery of the Abyss.* xiii+286.Vintage Books of Random House Publishers. New York, NY.

Light in the Darkness

Anderson, T.R. and T. Rice. 2000. Deserts on the sea floor: van Edward Forbes and his azoic hypothesis for a lifeless deep ocean. *Endeavour* 30(4): 131-137

Corfield, R. 2003. *The Silent Landscape. The Scientific Voyage of HMS Challenger.* xiv+285 pgs. Joseph Henry Press. Washington, DC.

Gilpin, D. 2007. *Spirit of the Ocean. Discover the beauty of our underwater world.* 256 pgs. Parragon Books, Bath, England.

Nouvian, C. 2007. *The Deep. The extraordinary creatures of the abyss.* 265 pgs. University of Chicago Press, Chicago, IL.

Rozwadowski, H. M. 2005. *Fathoming the Ocean. The Discovery and Exploration of the Deep Sea.* xii+276 pgs. Belknap Press of the Harvard University Press. Cambridge, MA.

INDEX

A
Agassiz, Louis, 32
Alvin Submersible, **31**, 163
Amberjack *Seriola spp.*, 158
American bison *Bison bison*, 142
American chestnut *Castanes dentata*, 52
American cowslip *Primula meadia*, 67
American crow *Corvus brachyrhynchos*, 92
American elm *Ulmus americana*, 52
American oystercatcher *Haematopus palliatus*, **21**, 110
American shad *Alosa sapidissima*, 132
American sycamore *Platannus occidentalis*, 67
Aristotle, 11
Atlantic bay scallop *Argopecten irradians*, **22**, 118
Atlantic cod *Gadus morhua*, 133
Atlantic mud crab, *Panopeus occidentalis*, **19**,105
Atlantic flyingfish *Cypselurum melanurus*, **30**, 159
Atlantic salmon *Salmo salar*, 132
Australian pine *Casuarina equisetifolia*, 74

B
Bachman, John, 31
Bachman's sparrow *Aimophila aestivalis*, 32
Bachman's Warbler *Vermivora bachmanii*, **3**, 31
Baird's sparrow *Ammodramus bairdii*, 31
Baird, Spencer Fullerton, 132
Bald cypress *Taxodium ascendens*, 67
Bald eagle *Haliaectus leucocephalus*, 106
Baltimore oriole *Icterus galbula*, 66
Banister, Rev. John, 23
Banks, Joseph, 20
Bartram, William, 8
Beebe, William, 160
Bellon, Pierre, 111
Black-faced Grassquit *Tiaris bicolor*, 146
Black locust *Robinia pseudoacacia*, 76
Black mangrove *Avicennia germinans*, **16**, 94
Black needle rush *Juncus roemerianus*, 88
Black oystercatcher *Haematopus bachmanii*, 32
Black sea bass *Centropristis striata*, **24**, 127
Blue crab *Callinectes sapidus*, **19**, 104

Blue grosbeak *Guiraca caerulea*, 23
Blumenbach, Johann Friedrich, 21
Bog laurel *Kalmia polifolia*, **9**, 62
Brackish-water fiddler crab *Uca minax*, 103
Brazilian Pepper-tree *Schinus terebinthifolius*, **13**, 77
Brisson, Mathurin Jacques, 21
Bristlemouth *Gonostoma elongatum*, **32**, 166
Brown-headed cowbird *Molothrus ater*, 52
Bufflehead *Bucephala albeola*, 74
Buffon, Georges-Louis Leclerc, Comte de, 21
Bullfrog *Lithobates catesbeiana*, 13

C
Caesalpinia pulcherrima, 8
California cord grass *Spartina foliosa*, 89
Calvert, George, 66
Cane *Arundinaria spp*, **3**, 34
Caribbean spiny lobster, *Panulirus argus*, **28**, 147
Carrion crow *Corvus corone*, 92
Catalpa sphinx moth *Ceratomia catalpa*, 53
Catesbya pesudomuraena, 13
Carolina jessamine *Geisemium sempervirens*, **7**, 56
Carolina parakeet *Conuropsis carolinensis*, 37
Carp *Cyprinus carpio*, 133
Catesbaea spp, 13
Catesby, Mark (1682-1749), 7
Catesby, Mark, quotes, 9, 10, 11, 14, 23, 24, 27, 28, 30, 36, 37, 42, 53, 60, 64, 67, 68, 70, 84, 92, 96, 97, 98, 100, 102, 108, 111, 112, 117, 121, 130, 143, 144, 154, 158
Cedar waxwing *Bombycilla cedrorum*, **14**, 82
Champlain, Samuel, 110
Chestnut blight *Cryphanectria parasitica*, 52
Cinchona officinalis, 8
Collinson, Peter, 7
Conch *Strombus gigas*, 152
Coneflower *Rudbeckia*, spp., 21
Cord-grass. See Spartina, **15**, 87
Cuban bullfinch *Melopyrrha nigra*, 143
Cuban grassquit *Tiaris canora*, 146
Cypress *Taxodium* spp, 30

D
Darwin, Erasmus, 20
Dolphinfish *Coryphaena hippurus*, 158
Dutch elm disease *Ophiostoma ulmia*, 74

E
Eastern black walnut *Juglans nigra*, 67
Eastern oyster *Crassostrea virginica*, **22**, 117
Eastern phoebe *Sayornis phoebe*, 56
Eel *Catesbya pesudomuraena*, 13
Empress tree *Paulownia tomentosa*, **12**, 75
Eurasian collared dove *Streptopelia decaocto*, 146
Eurasian crow *Corvus corone*, 107
Eurasian curlew *Numenius arquata*, 92
European green crab *Carcinus maenus*, 74
European starling *Sturnus vulgaris*, 74

F
False locust *Robinia pseudoacacia*, 67
Feral pigeon *Columba livia*, 92
Figwort *Scrophulariaceae spp*, 75
Filefish *Monacanthus ciliates*, 158
Fish crow *Corvus ossifragus*, **20**, 106
Forbes, Edward, 161
Franklinia *Franklinia alatamaha*, **6**, 46

G

Ginseng *Panax quinquefolius*, 52
Goldeneye *Bucephala clangula*, 74
Goldenrod *Solidago Canadensis*, 67
Goode, George Brown, 133
Gray, Asa, 82
Gray jay *Perisoreus canadensis*, 31
Gray triggerfish *Balistes capriscus*, **29**, 156
Graysby *Cephalophis cruentata*, **24**, 126
Gray heron *Ardea cinerea*, 97
Great blue heron *Adrea Herodias*, **17**, 96
Great hogfish *Lachnolaimus maximus*, 126
Greater Antilles bullfinch *Loxigilla violacea*, 146
Greater scaup *Aythya marila*, 74
Green heron *Butorides virescens*, **18**, 97
Green sea turtle *Chelonia mydas*, **26**, 142
Gum *Nyssa spp*, 30
Gypsy moth *Lymantria dispar*, 74

H

Halibut *Hippoglossus spp*, 133
Halley, Edmond, 17
Hawaiian crow *Corvus hawaiiensis*, 107
Hemlock woolly adelgid *Adelges piceae*, 74
Hooded warbler *Setophaga citrina*, 31
Horseshoe crab *Limulus polyphemus*, 153
House sparrow *Passer domesticus*, 73

I

Indian bean tree *Catalpa bignoniodes*, 52
Indigo bunting *Passerina cyanea*, 23
Ivory-billed woodpecker *Campephilus principalis*, **2**, 27

J

Jordan, David Starr, 132

K

Kentucky bluegrass *Poa pratensis*, 74
Knapweeds *Centaurea spp*, 74
Kudzu *Pueraria montana*, 74

L

Laurel-Tree of Carolina *M. grandiflora*, 23
Lawson, John, 8
Lesser scaup *Aythya affinis*, 74
Lined seahorse *Hippocampus erectus*, **30**, 158
Linnaeus, Carl, 13
Live oak, *Quercus virginiana*, 70
Loggerhead turtle *Caretta caretta*, 142
Long-leaf pine *Pinus palustris*, **8**, 57
Long-tailed duck *Clangula hyemalis*, 74

M

Maine lobster *Homarus americaus*, 147
Marlin *Makaira* spp, 156
Magnolia grandiflora, **1**, 23
Magnolia virginiana, 23
Marsh, O. C., 132
Martin, Maria Sibylla, 31
Menhaden *Brevoorta tyrannus*, 132
Mistletoe *Loranthacae spp*, 73
Morrow's honeysuckle *Lonicera morrowii*, **14**, 82

Mountain Laurel *Kalmina latifolia*, **9**, 62
Mud fiddler crab *Uca pugnax*, 103
Murray, Sir John, 157

N
Northern bobwhite *Colinus virginianus*, 146
Nassau grouper *Epinephelus striatus*, 152

O
Oak *Quercus spp*, 67
Olive nerite *Neritina reclivate*, **15**, 90
Orchard oriole *Icterus spurious*, 53
Oriental bittersweet *Celastrus orbiculatus*, 74
Osprey *Pandion haliaetus*, 106
Overcup oak *Quercus lyrata*, 30

P
Painted bunting *Passerina ciris*, 23
Pallas, Peter Simon, 22
Palo verde tree *Parkinsonia spp*, 71
Passenger pigeon *Ectopistes migratorius*, **4**, 36
Paulowna, Anna, 75
Pelicans *Pelecanus occidentalis*, 125
Periwinkle *Littoraria irrorata*, **15**, 89
Petiver, James, 8
Phragmites australis, 88
Pileated woodpecker *Dryocopus pileatus*, 70
Pine lily *Lilium catesbaei*, 13
Pine warbler *Setophaga pinus*, 31
Planthopper *Prokelisia marginata*, 89
Poinciana *Caesalpinia pulcherrima*, 8
Poison ivy *Toxicodendron radicans*, 77
Portuguese man O'war *Physalia physalis*, **30**, 158
Powell, John Wesley, 132
Purple loosestrife *Lythrum salicaria*, 83

Q
Quinine *Cinchona officinalis*, 8

R
Rain-lily *Zephyranthes atamasca*, 67
Rat snakes *Elaphe obsoleta*, 62
Ray, John, 12
Réaumur, René de, 21
Red-cockaded woodpecker *Picoides borealis*, **8**, 57

Red drum *Sciaenops ocellatus*, 137
Red-heart fungus *Phellinus pini*, 62
Red knot *Calidris canutus*, 153
Red maple *Acer rubrum*, 67
Red mulberry *Morus rubra*, 67
Red snapper *Lutjanus campechanus*, 126
Redwood *Sequoiadendron giganteum*, 66
Ribbed mussel *Geukensia demissa*, **15**, 90
Rice *Oryza sativa*, **5**, 40
Royal red shrimp *Pleoticus robustus*, **22**, 118
Rock pigeon *Columba livia*, 37
Rudbeck Jr., Olaus, 21
Ruddy turnstone *Arenaria interpres*, 153

S
Sabal palmetto *Sabal palmetto*, 57
Saltwort *Batis maritima*, 88
Sand fiddler crab *Uca pugilator*, **19**, 103
Sargasso crab *Portunus sayi*, **30**, 158
Sargasso frogfish *Histrio histrio*, **30**, 160
Sargasso sea, **30**, 157
Sargassum natans, **30**, 157
Sargassum fluitans, **30**, 157
Sargassum pipefish *Syngnathus pelagicus*, **30**, 158
Sargassum shrimp *Laender tenuicornis*, **30**, 158
Salt marsh, **15**, 87
Salt marsh periwinkle *Littoria irrorata*, **15**, 89
Saw palmetto *Serenoa repens*, 57
Scarlet ibis *Eudocimus ruber*, **16**, 89
Scup *Stenotomus chrysops*, 132
Sea mounts, **31**, 162
Sea oxeye *Borrichia frutescens*, 88
Shagbark hickory *Carya ovata*, 67
Sheep laurel *Kalmia angustifolia*, **9**, 63
Silk Snapper *Lutjanus vivanus*, **23**, 125
Sloane, Sir Hans, 8
Smooth-billed ani *Crotophagea ani*, 146
Sour orange *Citrus aurantium*, **27**, 146
Southern catalpa *Catalpa bignoniodes*, 53
Spartina alterniflora, **15**, 87
Spartina patens, 87
Stone crab *Menippe mercenaria*, **22**, 120
Striped bass *Morone saxatilis*, 133
Sumione oyster *Crassostrea ariakensis*, 123
Summer flounder *Paralichthys dentatus*, 137

Swainson's hawk *Buteo swainsonii*, 33
Swainson's warbler *Limnothlypis swainsonii*, 33
Swainson, William, 33
Sweet orange *Citrus sinensis*, 146
Sweetgum *Liquidambar styraciflua*, 30

T
Tanner, James, 27
Threadfin dragonfish *Echiostoma barbatum*, **32**, 166
Tilefish *Lopholatilus chamaeleonticeps*, 133
Tournefort, Joseph Pitton de, 17
Tradescant, John the Elder, 67
Tradescant, John the Younger, 67
Trichomaniasis *Trichomonas gallinae*, 37
Tulip tree *Liriodendron tulipifera*, **10**, 63
Turkey vulture *Cathartes aura*, 26

V
Virginia creeper *Parthenocissus quinquefolia*, 67

W
Water hickory *Carya aquatica*, 30
Water hyacinth *Eichhornia crassipes*, 77
Water oak *Quercus nigra*, 30
Whitefish *Coregomus clupeaformis*, 133
White ibis *Eudocimus albus*, **16**, 89
White-winged scoter *Melanitta fusca*, 66
White shrimp *Penaeus setiferus*, **22**, 118
Wild olive *Osmanthus americanus*, **11**, 70
Willow oak, *Quercus phellos*, 18
Withering, William, 20

Y
Yellow-eye snapper *Lutjanus vivanus*, 126
Yellow-fin tuna *Thunnus albacares*, **25**, 136
Yellow-rump warbler *Setophaga coronate*, 31
Yellow-throat warbler *Setophaga dominica*, 31

Z
Zebra mussel *Dreissena polymorpha*, 74

About the Authors

The interplay of art and science, combined with a penchant for travel steered the Brushes to create *Mark Catesby's Legacy*. M.J. worked professionally as a science illustrator and currently is a fine art painter. Alan is Professor Emeritus of Physiology and Neurobiology with broad interests in ornithology and the history of science. Both enjoyed careers at the University of Connecticut. Their interests and experiences combined to produce this exciting intellectual voyage through time and space.

M.J. and Alan invite readers to join them in contemplating Catesby's remarkable journey. They explore changes in the patterns and processes of nature, the interplay of life at myriad levels and some colorful characters from the past. This tale provides insights that enhance reader's appreciation of how nature works, from molecules to plants and animals, through entire ecosystems in this, our fragile planet.

CPSIA information can be obtained
at www.ICGtesting.com
Printed in the USA
LVHW051501220719
624870LV00021B/1989